UNSTOPPABLE MARKETS

FINANCIALIZATION, NEUTRALITY, AND THE LIMITS OF DECENTRALIZATION

NICHOLAS FETT

Published by Nicholas Fett

United States of America

First published 2026

ISBN 979-8-9963819-0-6 (Paperback)

Printed in the United States of America

CONTENTS

~

Bennet leaned against the hallway wall. It was his fault for arriving so early, but he had never been to this floor of the building. In his three years at the company, he'd never had a meeting with the boss, and he didn't want to miss it.

After ten minutes of waiting, the secretary finally opened the door.

Maybe he wants me to stay, he thought to himself as he walked in.

One foot in the door and he immediately knew that wasn't the case.

"I don't get it," said the older man from his chair.

Bennet quickly sat down in one of the seats opposite his desk. To him, Mr. DiPolenza was a celebrity. The plaque on the office and the building both bore his name. He was always here but rarely seen. To people of lower rank at the company, he was more of a spirit haunting the building. Luckily for Bennet, though, he was kind enough to meet the exiting employee, even if it was out of nothing but curiosity. After all, knowledge of new technology was supposedly his firm's expertise.

He stared at Bennet through his glasses, looking for an explanation.

Bennet cleared his throat. "It's like a normal futures and options exchange, but everything is done on a distributed ledger. The clearing, the settlement, and every backend operation is automated and performed by the smart contract I wrote. The parties don't need to trust each other at all."

"Sure, but who holds the funds?"

"No one has to hold the funds. The smart contract does," replied Bennet. He felt like a student who was proud to know the answer.

He knew he needed to calm down.

"Someone always holds the funds," chuckled Mr. DiPolenza. "And what happens when the operator of the smart contract runs into financial trouble and has to close down? Who's doing the reporting?"

"It actually, um, can't be shut down," Bennet muttered.

He took an audible breath. "Anyone can run it," he said a little more calmly. "It's decentralized, so you or I or anyone could copy the code and run the system. No one could stop us."

The older man just stared at the "whitepaper" on his ridiculously oversized desk. He'd been head of the trading firm for the past two decades and had seen his fair share of improvements for the financial sector. He'd also seen plenty of poor ideas from promising young traders. This pitch seemed like the latter.

He paused for what seemed like a whole minute.

Uncomfortable with silence, Bennet reached over and pointed to the diagram on the first page of his paper. He quickly leaned back down after an unwelcome glare.

Bennet continued, hoping to clarify. "It works just like the financial system does now, except software automates the entire process and can make sure the entire thing is solvent."

Mr. DiPolenza took a deep breath. He stared at Bennet differently. Not with a look of confusion or curiosity, but of pity. The way you'd look at a high school kid telling you he's going to propose to his girlfriend.

He said he didn't understand, but he felt as though he did. In his mind, the founder-to-be was the one who didn't quite understand.

He stood up and stared down at Bennet.

"Listen," he began, "your numbers were good here. You got a job any time you want one. I wish you the best of luck."

As Bennet stood up, he felt self-conscious of his suit. It was just more evidence he was outclassed. Maybe a nicer suit could have just put the meeting on more equal terms. At least in his mind, it might have helped.

“Thank you, sir,” replied Bennet automatically. He reached out to shake hands as if they were making a deal for something.

Bennet wasn’t offended at his boss’s lack of understanding. He told himself this was a sign of being early. He shouldn’t really expect these “TradFi” people to understand.

But it was still rough.

The meeting went sort of as expected. He still wished he didn’t feel as much like a child. Like an eight grader telling his dad about his idea to start a landscaping business, any support seemed backhanded and almost derogatory.

As he walked out, he remembered he had forgotten to make eye contact.

Off to grab his box of things and meet his fiancée for lunch before heading home to what would be his new office.

INTRODUCTION

"Acceptance is usually more of a matter of fatigue than anything else."

-David Foster Wallace

Markets are a method of distribution.

Whether it's who gets to live in a house, how many TVs should be produced, or who gets issued newly minted cryptocurrencies, the goal of markets is to remove human input from the decision of who gets what. They give distributional decisions to a process. The result is the sum of repeated neutral tradeoffs.

They can be used to help coordinate human action and efficiently put things in the hands of those who want them most. At the same time, history has shown they can be a driver of social disruption and extraction.

The core thesis of this book is that neutrality and financialization are failures.

Markets are increasingly being adopted so people do not have to make the hard decisions regarding distributions. By moving to an emotionless or neutral system, they remove the human element in deciding what "fair" means. The acceptance of any outcome and the neutrality towards its participants make these systems capturable and useless to a functioning community.

Financialization is the introduction or increased exposure of things to markets, be it through increasing liquidity, transferability, or digitization. It can allow parties who do not trust one another to interact, but it can also erode trust and longstanding social norms when introduced to previously nonfinancial interactions.

Cryptocurrency and digital assets have become synonymous with these two concepts.

I made my way into cryptocurrencies around 2011, just a few years after Bitcoin had launched. At the time, I was an economist who, like many, had been taken by the libertarian moment of the post-financial crisis. I was convinced of the power of the market and its ability to work for everyone who put in effort.

Naturally, I was drawn to Bitcoin.

Pitched as an experiment in overthrowing the government's monopoly on money, it represented a counterculture tech movement to escape the hyperinflationary time bomb that was the Federal Reserve and government-controlled money.

Like many early community members, I was convinced that the national debt was unsustainable, prices would rise as a result of its monetization, and society as we know it would turn to alternatives. Bitcoin was that solution.

Those of us who felt that borrowed or printed money was flowing to undeserving individuals were smitten by this new digital system. The rules and unforgiving automation were not handled by traditional legal contracts or relationships but rather code, cryptography, and markets. The original story was that anyone could hold and spend Bitcoin anonymously and instantly across the globe. "Like a credit card, but anonymous and free," I used to tell newcomers.

It sounds amazing, but looking back, it's hard to say the cause was even justified. There was no inflation. There was no imminent implosion of the traditional system that made this necessary. Despite significant fluctuations in reported deficits and monetary aggregates, the oft-predicted hyperinflation never occurred. And yet, for over a decade, people have continued to try and sell crypto as a monetary system that's not being devalued.

Regardless of the original intentions, Bitcoin was just the beginning, and we've created countless systems since. Financial markets specifically are transforming. Legal contracts are being redesigned to have automated and open-source resolution. Information itself is being financialized, and its validity given to the world of prices. Human interactions are being separated from the intermediaries and institutions whose necessity is enforced by regulations.

This is the promise of decentralization.

We're replacing the social elements of government, the custodian, and the intermediary with an unstoppable piece of code. Every asset is turned into a digital representation of itself so it can be financialized and given to a market.

Cryptocurrencies as money and finance represent but the first of many "unstoppable" codifications of our society. Your wallet and digital life will come first, but how companies are run, how goods are sold, and how governments operate are next. If structured properly, they can be used to steer society towards meaningful ends, and they can give us new options for organizing businesses and communities.

Unfortunately, these markets and systems are not quite there. If anything, they are completely failing to do what they set out to do.

The details matter.

These markets, like all markets, are planned and designed. They are systems we create, not natural laws we discover. As such, they should be judged and adjusted based on whether they improve the broader society they exist within.

Bitcoin was just the first of many experiments along these lines. From Ethereum, proof-of-stake, endless tokens with unfathomable liquidity, decentralized exchanges, prediction markets, stablecoins, and many more, we're playing around with the base coordination tools of a new society, and we're finding out it's really, really hard.

Markets and financial innovations can create unbelievable economic and financial gains. They can also result in outcomes that are far from desirable. Their implementation is a choice, and success depends on how they are designed and maintained. If crypto is used to push back against the captured institutions of the physical world, it can help. If immutable financial markets are looked at as the final solution, society will undoubtedly end up worse than before.

~

Put your head down and build. Create a vision and a purpose. Iterate ruthlessly until you find product-market fit. Be everywhere at once.

To Bennet, being a founder felt like trying to capture the wind. The constant grasping and movement with little to show for it. He had been doing this for a few months and was still getting the hang of it.

Apparently, there was a process to follow, and he did his best to follow the advice of the talking heads and success stories. He'd watched more YouTube videos on running a startup than his previous self would have ever guessed existed. In some ways he felt like an expert, but practicing the art was still a little awkward.

Ready for the challenge, he knew one of his first tangible steps would be to get feedback on his system.

Only knowing a few people in the crypto ecosystem, he decided to hop on the Amtrak up to the city to attend FreedomHack, an event for developers in the community to apparently hang out and build software together.

The "hackathon" was a new concept to him.

He learned coding over the past few years and did a little in undergrad, but going to write code for a weekend in order to compete for prizes wasn't necessarily his first idea of a good time.

Bennet was pleasantly surprised when he arrived. Full of college students and a few "older" folks like himself in their late 20s, they gathered in what seemed to be an old warehouse in Brooklyn for the event. Just tables, cords, and a few sponsor booths around the outside; there wasn't a suit in sight. The few other 'hackers' who had also arrived on time were sitting with headphones on, clearly working diligently on their stickered laptops. The atmosphere didn't bother Bennet in the least.

Despite no one saying a word to him, it felt welcoming. The vibe was definitely one of being early to the space. He noted for next time that his "casual" look of a sweater might still be a bit too formal.

After walking around to the booths of the successful projects, he went to the team formation area, a small stage in the back of the giant room.

There were a handful of presentations from some first-timers looking for a team and a few other "founders" looking to create tokenized art or digital ownership of music rights. He quickly realized that his financial ideas and economics-heavy proposition would need a hard left tweak to capture some respect from this crowd.

He gave his presentation on his new chain for decentralized finance. He focused on the why. The pure decentralization. The vision to kick the government out of anything to do with finance. The libertarian dream of unregulated financial markets.

It really wasn't an idea for a hackathon, but to his surprise, he got a few people on board. Granted, they were just looking for a teammate that seemed competent and could write some basic smart contracts, but he'd take it. One was Annika, a girl from Berlin who was wondering if they could use Bennet's structure to fund donations for anti-war efforts. The other was Josh, a devOps engineer who wanted to run backend stuff and was clearly just looking for someone with a vision to latch on to.

Not really knowing how these things worked, they made a team, and the idea for the hackathon was Annika's donation cause.

Considering Bennet's chain wasn't live yet, they couldn't really do much regarding his vision. He decided to just build on some of the other platforms to check out the competition. To his surprise, it didn't seem like anyone had working software besides the ability to launch a token. After working diligently throughout the next two days, they too produced a similar product. In the end, they presented a new token with a website that looked like they didn't know how to build websites.

All said and done, though, they got third place thanks to Annika's idea being right on target for the judges.

Bennet reached out to Josh after the hackathon to see if he wanted to continue working on the chain. Josh still had a full-time job, so Bennet was relieved to find some help that he didn't have to pay for. He had thought about bringing Annika on, but Bennet figured Annika had too many of her own ideas to ever be an employee. Also, being recently engaged, the idea of working too closely with a female made him a little apprehensive.

1

DEFINING MARKETS

"The ultimate, hidden truth of the world is that it is something that we make and could just as easily make differently."
– David Graeber

The year was 1889.

Over 50,000 people lined up along the starting line on the Oklahoma border. Horses, carriages, and runners, all carrying flags. For weeks, these predominantly white men and their relatives had been camping in shoddy tents along the railroads that represented the line between American and Indian territories.

Today was their day.

At the sound of a cannon, they raced to lay claim to their plot. Pushing across the 1.9 million acres that was now open to settlement, each man would ride to his desired location to plant his flag, securing his own 160-acre plot of land. He would then race to the nearest land office to get it officially registered, after which he would retrieve his family to begin building their new life.

Disputes were common, and the entire process took years to sort out the competing claims, but the process was generally looked at as successful. Over the next decade, land runs became a common way to distribute the "unassigned lands" that were given to Native Americans over the previous centuries.

The Oklahoma Land Run of April 22, 1889. Photograph ca. 1889. Courtesy of the Oklahoma Historical Society

STEPPING BACK, this wasn't something that happened overnight. America was a little different in the early 1800s.[i] Land wasn't looked at the same way as it is today, and there were quite a few competing views on what should be done with the land.

For most of the previous century, many of the early farmers who'd ventured off the East Coast were basically squatters. They were typically poor settlers who just found some open land in the vastness that was the US, and then they started farming. They didn't necessarily own the land or even try to. In 1807, though, there was some pushback from people now looking to use that land. Congress responded and passed the

Intrusion Act, which made illegal settlement punishable by jail time and authorized the president to use military force against squatters. New England congressmen criticized the bill as tyranny and "bayonet justice" that denied constitutional rights to these settlers simply because of their class.[ii]

There were two main issues at play around public lands: preemption and distribution. Preemption allowed squatters the right to the land they settled and improved prior to any public sale, while distribution involved the question of what to do with the proceeds of the land sales. Congress approved some smaller preemption laws from 1830 to 1840 that helped settlers who were already squatting on land, but westerners wanted those rights to extend permanently into the future. At the same time, Republicans, and later the Whigs, fought hard for a distribution bill that would allocate any surplus proceeds from public land sales to the states as opposed to just the omnibus federal budget.

There was other land to be distributed too: the native lands. This land was theoretically free from squatters, but it was now being debated as to how it should be split up. On the one side were the northerners who had long fought for a program of land grants. They were in familiar opposition to the slave owners of the South who, generally speaking, wanted to sell parceled land to the highest bidder.

The "Free Soilers" of the north believed that land should be a given right for families. The idea of the "yeoman farmer," or in modern speak, the self-sufficient homesteader, was alive and well in mid-1800s America. Sixty years prior, the US even passed a law that limited the maximum-size lot one individual could purchase of public land to 160 acres. They believed that land should be given to individuals and families rather than the wealthy plantation owners who would develop the land using slaves and cheap labor.

In addition to the rich who owned farmland, another group

that was against free land was the employers up north, namely the factory owners. They saw the ability of employees to leave for land out west as a direct threat to their labor markets. Henry George, a famous economist of the time, wrote on how the ability of workers to just leave employment to start their own farms was actually the best way to keep labor practices and wages fair.[iii]

After decades of debate, the start of the Civil War marked a period where the Northern government didn't have to listen to the Democrats of the South and their leader, President James Buchanan, who'd left office a year prior (1861).

As a result, and to the disbelief of contemporary cynics, the American government chose the path that defied the powerful lobby of industry and wealthy landowners, selecting instead a distributive model that favored individuals over capital. The first of many, land runs, land grants, and homesteading became a popular practice, and the distribution of land out west looks much different than it would otherwise.

Between 1862 and 1934, some 270 million acres were granted by the US government. Homesteading in the US lasted for over a century longer, with the last state being Alaska, ending the practice in 1986.[iv]

Markets are a choice.

They are a form of distribution through auction. If there is land to be distributed, you should simply sell it in whole or in parts to the highest bidder. If we do this with every good or service in our economy, we live in a "free market society." If you want to start a company, distribute land, labor, tokens, or even database entries in a ledger, the "market solution" is to give the product to the person willing to pay the most with as few constraints as possible.

They work because rather than focus on a collective agreement, distribution decisions are compressed down to individual tradeoffs. There's no more problem of democracy, voting, or consensus. The final distribution is clearly what society revealed as its preference.

This seems relatively straightforward, but the unintended by-product is that the distribution is now not the decision of society, but rather the aggregated result of each individual's decision.

The classic example in economics textbooks is that of barter. If a given community has some cows, chickens, and bread, how do they distribute the items fairly?

Compared to a top-down solution where a leader might dictate everyone gets a cow, three chickens, and a loaf of bread, a market system will allow individual preferences to determine the distribution. Maybe you like bread more than meat. Maybe it's really good bread. Either way, instead of distributing a good equally or in a way that has some historical precedence, each good is given to the person willing to trade the most for it.

Assuming no regulations, society is now "neutral" toward the distribution. The decision has been handed to the market. If one person wants to buy everything, he has every right to. If one person owns more at the start of the market, he has an advantage. If everyone with resources wants one animal and the other becomes no longer available, market proponents would claim this is clearly what the people want.

But markets haven't always been as pervasive as they are now. In fact, even the historical views regarding ownership and distributions were not as straightforward or neutral as they are today. The purely "free" markets of the barter sort were reserved for those with whom trust and social coordination were difficult.

And it's for this reason that markets became the norm. As

our societies grew and technology increased, we had more people to interact with.

Early on, most of civilization ran without money or even the need for long-distance interactions. In fact, besides a few small exceptions in major cities (Egypt, Athens, etc.), life for the average human was that of a peasant or small-scale farmer. You had social interactions with those close to you, largely governed by a local leader or other theocratically ordained person of power. What you did in life was determined by what your parents did in life, and how you distributed any assets was determined by, again, what your parents did. Even besides the fact that there was no meaningful excess, people stayed in one area for their entire life. You watched out for those around you and shunned and feared that which you didn't understand.

As is the case with small groups now, distribution was largely communal, religious, or tradition-based. If you think about how families distribute food or resources, it's definitely not a market-based system.

And don't be confused or give them any credit; this wasn't really a conscious choice. They couldn't really communicate with or enforce things that were too far away. The technology was just too limited at the time. Even things like governments and military decisions were local and limited in scope because a ruler literally couldn't monitor or communicate with his officials. You can think back to Athenian democracy or Sparta; they were each just one city.

And this is how most societies operated for long periods of time. You used barter with enemies and far-away trading partners, but close to home, things were relatively stable. Markets existed, but things like land, labor, and education were done differently. Tradition and social norms held roles that limited the pervasiveness of market forces.

Over time, things gradually changed. Ships got faster, the printing press allowed information to spread with a universal reach, and the telegram enabled communication in real-time. Information began to permeate across the globe. Ideas of economics, politics, and science were now no longer limited by word-of-mouth transmission. The world grew rapidly. Legal documents were copied and enforced. Land titles, accounting systems, stock certificates, and more.

It's around this time that markets began to accelerate. The old ideas that you would just trust your neighbor or those close to you were limiting. How do we interact with those in faraway cities who we can't trust? How do we handle these larger, sprawling cities? The only answer is that of barter markets and mutual trade. We had long used these systems with distant countries (e.g., the Far East) or some enemy tribe, but this time it was different. As the cities grew to account for increased industrialization, markets began to be used even near home. The land you farm, the workers you employ, or even food itself became staples you have to earn and buy at market rates. It's essential to note just how foreign and radical these ideas were.

This was a huge cultural shift.

On a historical timeline, this method of economic organization is brand new. That said, we're still trying to answer the same questions. How do we distribute the food or land in a fair manner? How do we spend our time productively (labor), and what should people be paid? How do we organize risky ventures, form businesses, protect communities, or incentivize cooperation for public goods?

These questions we're trying to solve with our innovative tech companies are the same basic problems they had 500 years ago. The difference is that these decisions that we once made as a society or through tradition are increasingly being handed to the market.

Just as in our land grant story, there's never a blank canvas. The Native Americans were the clear losers, but even beyond that, is distributing the best land by who's the fastest rider really a fair method? Did they give enough notice to everyone to be on the starting line in time? If they had chosen the auction route, distributions in wealth would clearly make the auction biased (the landowners and industry have the advantage). But what if the money raised by the government on the land sales was put to good use? Did we pick these free-soilers over other citizens? All market creations cause disruptions and pick favorites, even if not as blatantly as stealing from natives.

Another example is that of labor markets.

Prior to the 19th century, most of Europe operated on a system of serfdom, with the majority of people's lot in life chosen by the place and manner of their birth. Goods, land, and labor were distributed largely by tradition. Serfs were bound to the land and owed labor services to their lord in exchange for the right to farm strips of land for their own families. Over the years, however, laws were gradually introduced to remove serfdom and universal wages (standard wages for "free laborers").[v] Populations were soon subject to competitive wage arrangements.

The overall effect was the pushing of farmers into cities to work for industry. Economic theory would claim that if the factory pays the most, it makes sense that society should push people to work there. Unfortunately, as many other countries realized, moving workers away from their families and towards the city has profound social consequences. Even if the wage were higher, the free movement of labor to satisfy the needs of the market changed the entire society. The once communal landscape of serfdom was transformed into a competition for livelihood, and many lives and social structures were completely upended as a result of this transition.

The workers learned the hard way that the market destroys old industry as it creates new ones.

modernity

The story so far is that as technology improved, we could communicate more with people far outside our own personal region of trust. Therefore, we began to use markets and bartering with greater and greater frequency. Although this built financial wealth, it often came at the cost of social disruptions.

The trend has continued throughout the past century. The printing press, the telephone, and the internet all allow us to interact across larger networks. Anyone can now communicate instantly across the globe. The only thing we couldn't do is send the actual good or value we're bartering for.

Crypto is this final stage of technology necessary for instantaneous global markets.

Digital assets, like crypto, allow us to measure and transact with value from the physical world. Each token is, in essence, a representation of something that we give a medium to. If we move these numbers on a screen, they affect real-world distributions and activity. In other words, you can trade them for things or pay people with them. If we all pretend that this digital certificate is the real thing, we can now send it around the world, trade it more efficiently, or automate things we couldn't have.

Imagine we create 100 tokens that represent a piece of land on the outskirts of a city. We can then trade around these tokens for ownership rights of the land. It could allow for better price discovery. Maybe some buyers who couldn't visit it can participate. Maybe some speculators begin to take part in the trading. Maybe buying and selling just got much cheaper than signing titles and physically going to an office. In any case, the land is just land; it hasn't changed, but the creation of a digital asset and the market for these digital goods change how we talk about things like ownership or especially who has rights to farm or build on the land.

And this works whether there is land or not. If I make a token and someone is willing to give me money for it, that token facilitated a real-world transfer of value.

This financialization, the increased exposure to market forces via digitization, is crypto's main use case.

And make no mistake, markets do help when we can't trust one another. Even within our own communities and governments, people succumb to mob rule. Leaders can often be harsh, if not cruel, to minority rights. Institutions that traditionally stood to organize society can be captured.

Automation, markets, and crypto can remove these forms of capture by moving interactions to a more neutral system. They can prevent government favoritism and can incentivize economic activity through competition. These are real benefits, and the adoption of markets can help if the resulting distributions fit with expectations.

The problems arise, however, when they don't. Like the institutions they seek to replace, capture is seemingly inevitable. We're realizing over the years that markets and automated systems don't always result in optimal outcomes. Society finds itself constantly pushing back against those seeking to solidify their standing resulting from the current method.

But distributional decisions, like value, are a social activity. Opting into the removal of this ability via automation and code is a choice, but it's also a choice that society can opt out of at any point.

In the case of land runs, the government decided to distribute land without a market. First come, first serve after the starting gun. It's not a method that governments would even consider now, but the important lesson is that there are options.

As we've seen with other market introductions throughout

time, societal views toward existing distributions can and should change. The system and its outcome are choices to be made by those who interact with it. Crypto is predictably discovering that they have more choices to make than they'd like to admit.

market warlords

A warlord sails to a nearby island.

He hops up onto the mast of his large battleship and proclaims, "I'm going to kill all of your men and enslave your women and children."

Luckily, he continues. "This will happen after my vacation... and the even better news is I've decided to make 'safety vouchers' that we're having an auction for. You can bid on them, maybe trade them around, and if you hold them when I get back, it will let your family be safe. See you soon!"

The island was stunned.

"What luck," they claimed. "We're so fortunate that the warlord understands markets."

The islanders quickly bid on the vouchers. The vast majority of the vouchers go to the old rich people on the island who have gold (the currency the auction is in). Some trading happens after the fact, but it largely stays with the wealthy.

A few days later, one of the islanders asks if they can just distribute the vouchers to the young kids who can keep the island running for the next few generations. The economist on the island then refutes, "Clearly you don't know economics; a free market is the only way to distribute things efficiently."

~

"Dojima is the chain for decentralized finance," said Bennet as he pulled up his slides.

"The derivatives industry is worth five trillion dollars... and we believe this is too big."

He paused for a smile. This was his zinger that was supposed to grab their attention.

"We're going to drastically reduce this number, but we're going to own the entire thing. By decentralizing every aspect of the industry, we can drastically reduce costs and extraction from inefficient middlemen. From user-controlled funds, automated settlement, and true censorship resistance, we can enable everyone in the entire world access to the financial tools once reserved for Wall Street."

For the first time in his professional life, Bennet actually believed what he was selling. He'd practiced this speech more times than he could count over the past few weeks. He wanted more than anything to raise money to prove not only his vision but also his ability to do this crazy idea of starting a business.

He'd given quite a few other pitches over the past few months. This time he was lucky enough to be in person.

Of the two VCs he was presenting to, one was a tall Asian man in a Patagonia vest and highly shined dress shoes. He was supposedly a big deal in the space. The other was a semi-attractive Asian girl who Bennet was thankful he could rest his eyes on during his routine.

Dojima was taking part in a crypto accelerator in exchange for 10% of the company.

After realizing that starting capital was hard to come by, Bennet agreed to join, as they promised to introduce him to the best venture funds in the space. These were the final days of the eight-week program, and at least in Bennet's mind, it was definitely worthwhile.

He hadn't gotten any takers yet on the pitch, but even getting people's time was a luxury after months of cold emails. The VC firm they were currently talking to had connections with the largest exchanges, and lots of the bigger token projects were in their portfolio.

He'd practiced this pitch hundreds of times to make sure it was perfect.

After the 3-minute presentation, the investors quickly started on their questions.

"Can you tell us about your marketing plans?" asked the girl.

"We plan on growing organically. We run the crypto meetup in our city, and we've been growing organically on Twitter and reddit too. The technology is really just something a step above, and we plan on focusing on the product," replied Bennet.

The two investors were writing as he answered.

"What connections do you have with exchanges?" she asked quickly. "And are you in partnership with any market makers for the launch of your token?"

"God damn, shit questions," thought Bennet. But he had a prepared answer. "We feel that exchanges will come organically. Also, we're part of this accelerator and also have experience building decentralized trading tools, which we plan on utilizing."

The three seconds of silence showed they clearly weren't impressed. Bennet felt like he needed to stop saying "organically".

"What is the token vesting schedule?"

This meant how soon could they cash out.

"No vesting; if you want out, you can get out," said Bennet. "We take Satoshi's vision to heart. This isn't some system that's going to be held together by off-chain legal contracts."

The investors perked up with that answer.

The girl put her pen down. "Can you tell us the size of your community?" she said, finally meeting eyes with Bennet.

She was clearly asking about social media numbers. Bennet didn't really have an answer there.

"Yeah, we have a large discord community and participate in AMAs on telegram frequently. The community is very engaged, and they're super excited for mainnet launch." He stretched the truth but figured they could due diligence him if they really cared.

The male investor wrote for five seconds and then picked up his head. The girl gave him a nod. They thanked Bennet and Josh for their time and gestured toward the door, expecting the two to send in the next team.

Bennet was slightly disappointed. He didn't even get to show off the application or explain any of the technology. The design of the system was a thing of beauty to him, but this was the third firm that didn't even ask. Of those that did, it seemed to be more of just a personal interest in the tech rather than any real investment strategy.

Later that night they received an email with a request for a term sheet along the lines of giving the investor five percent of the tokens at launch. The deal was contingent on giving another ten percent to an Asian exchange for the purpose of listing and market-making the token.

Since it was 2019 and few other investors had given them the time of day, Bennet decided to go for it.

2

STARTING THE CHAIN

"You've got that eternal idiotic idea that if anarchy came it would come from the poor. Why should it? The poor have been rebels, but they have never been anarchists; they have more interest than anyone else in there being some decent government. The poor man really has a stake in the country. The rich man hasn't; he can go away to New Guinea in a yacht. The poor have sometimes objected to being governed badly; the rich have always objected to being governed at all. Aristocrats were always anarchists."

– G.K. Chesterton, The Man Who Was Thursday: A Nightmare

Once upon a time, there was a Native American village in which the people lived peacefully, spending their days hunting and gathering. They shared everything, distributing to one another what each person needed. As they started getting bigger, however, some people felt that they did more work than others. A few of the bigger young men were breaking their backs hunting and carrying elk to the village, while the older folks and women and children had the easy job, or so they claimed, of picking some berries and watching the kids.

Seeing the validity in their claim, the chief, an old man who liked building things, decided to come up with a solution. He worked all night and, in the morning, presented to the tribe a gigantic wooden wheel with slots on it for each person. At the end of each day, each person would put everything they picked or hunted into their family's slot. Powered by the river, the wheel would then raise the sides up and spin for several minutes, pushing all of the goods to the middle of the contraption. Then, somewhat magically, the machine distributed everything based on how much each person put in. If there were 100 pounds of food total, and one person put in 10 pounds, he would get 10% of each item. It was a truly miraculous device. He dubbed it the "merit-go-round."

Skeptical at first, the villagers tried it, and to their collective amazement, it worked. The people who put in the most seemed to get the most, and those that didn't work that day didn't get any. Our original hunters were thrilled. Finally, they had their fair share.

Unfortunately for the system, though, people quickly realized that it wasn't just the contribution amount but rather the weight. It wasn't exactly work that paid, but rather just that you managed to go get the heaviest item available.

As a result, the citizens started providing the contraption with heavier things. After the first month, nearly everyone had given up picking berries and was instead hunting and fishing. After the second month, most people started to soak all of their items in water for a day before putting them in the machine. Most of the meat was therefore pretty awful after that.

It also became pretty evident that there were some freeloaders on the system. Mothers with young children, the elderly, and those that took time out of their day to care for them didn't have nearly as much to put into the machine. Since the villagers never knew when they might need to take time off, everyone started storing and saving. Whereas in the past they would just share it all, now each tent had a large storehouse for dried meat out front. The bear problem was quite a sight to see after this introduction.

Some of the larger families even decided to quietly opt out of the system. Sick of eating wet meat, they decided to just distribute things privately and let other members pick berries that they too would keep private. Calls for enforcement came to "encourage" contributions of the good items.

A few months into the experiment, the machine was mysteriously destroyed, and the tribe was happily forced to go back to the old system.

A MARKET OPTIMIZING for some non-ideal factor is nothing new. Throughout history, market structures are introduced to a society in hopes of finding cheap goods or quality services, and they get something unexpected. The solution to distribute things fairly turns into a game that no one wants to play.

We see it with housing in the US. The market structure for mortgages has individuals promising ever-increasing portions of their future earnings.

We see it with natural resources. Supposedly free markets are captured by geopolitics, and the winners turn out to not be those that can produce, but rather those that can win at the competition of military industrialism.

We see it with social media. Monetization of likes and views was supposed to produce entertaining and original content. Instead, it promoted slop and reposted content.

There are real-world constraints, and market designs need to account for them or adjust as they occur.

Crypto is learning this lesson the hard way.

crypto as the contraption

As with our young hunters looking to escape a seemingly unfair distribution, Bitcoin was started with similar ideals. The founders sought to replace the distribution and control of money.

To do this, they created blockchains. A blockchain, at its core, is conceptually similar to that of a standard spreadsheet. A technology for keeping track of data. To simplify, when thinking about these systems, think "databases."

The general idea is that digital money at its core is nothing more than a ledger of who owns what. Therefore, to create a digital money with no one in charge, Bitcoin created a database that could perform the task.

Whether it's a traditional database or a blockchain, the hard part of maintaining a trustworthy system is choosing what the rules are for adding new data and storing old data. At the end of the day, a database is a program running on a physical computer. Who can add new data or change the history and also who stores and maintains the computers are all decisions that need to be made.

Traditionally, one company does all of the above. They use a centralized model similar to just having a spreadsheet on your own computer; it's their resources doing every piece, and they can change and delete what they want. If the company has a spreadsheet of bank balances or personal data, they have all the power. They control the database and can literally change anything they want or delete the whole thing if they, or the main concern of crypto enthusiasts, the government, want them to.

The problem with traditional databases is that they require trust.

That owner isn't neutral.

You often can't see what they type into the database, and

you have little recourse if they make a mistake. For the cypherpunks of early Bitcoin, they didn't want to have to trust anyone. They hated the Federal Reserve and its post-financial crisis bailouts. They wanted a database for a currency that couldn't be changed by some money-printing oligarch.

The novel concept of blockchains is that instead of just one database operated and updated by a centralized party, we all have a copy of the database and then have rules for updating it. If we all hold a copy and know the rules which must be followed, the system is secure. There is no one party that can update or change the history on the users. When we want to interact, we check that our databases are synced up, and we should be good to go.

These rules for how things can be added to the database can be simple or complex. A naïve approach might be to have everyone who has a copy of the database take turns making the updates (e.g., we just go round-robin). The problem with doing it this way is how to maintain anonymity and open access (who gets to pick who's in the round-robin list). What if someone claims to be two different people? Or 100?

For Bitcoin, the solution was proof-of-work (PoW).

Proof-of-work is the competition for who gets to update the database next. It's the solution to the problem of who gets to order the transactions and update the database.

The system works by using a cryptographic puzzle where the solver needs to produce a hash with at least a certain number of leading zeros. To simplify, you have to guess the correct number faster than anyone else.

This process is called "mining."

Anyone wishing to update the blockchain will take the database (the current state + any new additions) and then try to guess a random number to solve the puzzle. The first person to solve it wins and gets to pick the list of transactions, including a reward of several Bitcoin for themselves.

This "mining reward" is crucial.

It's the innovative market design of Bitcoin. It incentivizes people to compete to update the database and check that the updates follow certain rules. It's a clever system, and all it requires us to do is verify that the miner solved the puzzle.

The Bitcoin system for updating the database is a market.

Most people know that Bitcoin is an asset that you can use as a currency, but the asset isn't the market. The market created is the act of the system "selling" new Bitcoins for a correct solution to the puzzle.

Since the creators wanted no external reliance (e.g., connection to the US dollar), they needed a way to do an auction with no money. This is where proof-of-work is unique. It uses cryptography to necessitate spending on hash power (e.g., electricity and compute). This hash power spending identifies the highest bidder.

In some ways, the mining competition can be thought of as a lottery. Parties buy specialized hardware and spend money on electricity to enter the lottery (to guess the random number). If you pay more, you get more tickets. If someone wants to buy up all the tickets to censor the network, they'll likely need to spend more than everyone else at every block (once every ten minutes). It would quickly get expensive. Thinking of it this way, PoW is an auction for new Bitcoin with the externality that it secures the database. The market didn't randomly appear; it was created and designed with a specific purpose.

goals of crypto

In order to say whether Bitcoin was well-designed or is functioning properly, there needs to be some objective or goal to judge the system against. Of course, the metric of the price of Bitcoin has been wildly successful. For other measurements, though, it's not quite as clear.

The goals of Bitcoin were largely rooted and expressed through the desires for "decentralization." Early proponents needed a secure database for a currency that wasn't controlled by anyone, a system where no one could censor, change the rules, or stop others from participating.

As per the title of the whitepaper, "peer-to-peer electronic cash," the selling point was that there wasn't an intermediary. The early pitch described it like a credit card, except it was anonymous, free, and no one could stop the transaction from going through. Of course, for those who bought early, there was the added benefit that it would be worth more money in the future once people realized how awesome this technology was, but that was just an added bonus. Its primary purpose and utility centered on its use as a currency that no one could stop or print more of.

What the early founders meant by decentralization is still debated today. No one is quite sure which piece actually matters.

Does decentralization simply mean we can all see the Bitcoin database?

Or does it mean we can all validate it or participate in mining meaningfully?

Does it mean we can self-custody (not rely on a bank or exchange to hold our money)?

Does it matter if we actually do?

What about spending it and who accepts it as a currency?

Do I have to be allowed to spend it, or is that a secondary issue?

What if I only have a tiny amount and don't have enough to pay high fees? Do I count?

There are no correct answers.

The most accepted version of decentralization is that you can run a node (verify the changes in the database), the validator set can be anyone (no permissions on mining), and no one can unanimously delete people's balances or change how the system works. It seems simple, but even these small goals elude most cryptocurrencies.

Despite the confusion, the task for Bitcoin was straightforward; the system needs lots of people running nodes (holding copies of the database) and lots of people competing to add new updates (miners). If the first is compromised, then one party could change the history or the rules, and no one would notice. The second is important because whoever gets to make the update chooses who gets to move their Bitcoins. If you control every update, you can effectively prevent people from accessing their own money. Therefore, if you want to have claims of censorship resistance and consistent rules, you need "decentralization."

personal computers

To achieve this dream, Bitcoin started as a system where anyone could operate mining software on their own personal computer. The dream of many in the community was that everyone who owned Bitcoin would run a small program on their laptop to collectively maintain the blockchain while earning modest, distributed rewards.

Unfortunately, this original idea was almost immediately thrown out the window. There was money to be made, so economies of scale kicked into action. People realized that they could get more rewards if they started writing custom Bitcoin miners. First, they utilized GPUs (the graphics cards on your computer) for faster processing. Then companies took it further and started making FPGAs (programmable chips) to perform the task. After a while, even ASICs were made for Bitcoin. These extremely custom chips do one thing only and are the pinnacle of fast computation.

There is nothing wrong with innovation and competition,

but the problem became that whoever had access to the newest and best hardware could outcompete and monopolize the market (something very anti-decentralization). If you developed these new methods faster, you could win the PoW competition much more often than the other miners.

People quickly realized that the market for new Bitcoins was dependent upon two main factors: mining hardware and the cost of running it (electricity and security of the machines). As a result, the system centralized along these two lines.

One company in particular, Bitmain, became the premier research organization for developing new ASICs, and they continue to control a large portion of the network because of it.

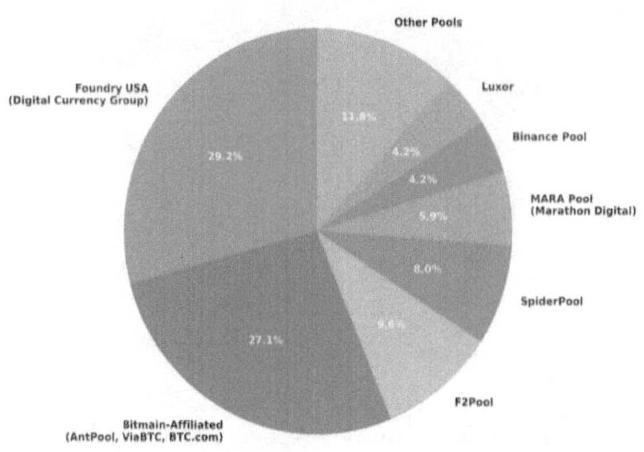

Bitcoin Mining Distribution, Aggregated by Controlling Entity. Author's own figure.

As of January 2026, two entities control over half of the network's hash rate.[i]

access to energy and security

In addition to developing the hardware, the location of the mining equipment is just as important. Of course, one should want the place in the world with the cheapest electricity, but unfortunately if the network consumes as much power as Poland (which it does), it's likely not going to be a small country, and in fact, the miners might even need to build new electricity sources (which is happening).

The mining hardware also needs to be secured. Tens of billions of dollars' worth of physical hardware can't just be in any country. It has to be a very secure country, and they either have to be in on the operation or very willing to allow the operation, or there's a serious risk to the investment. And this is exactly what we see.

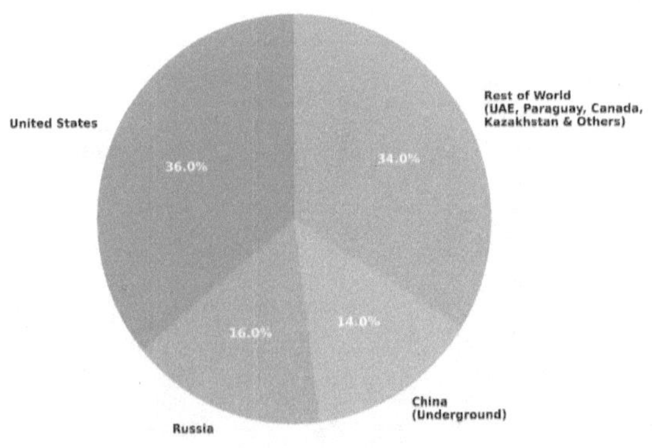

Bitcoin Mining Distribution, Aggregated by Geographic Concentration. Author's own figure.

The US, China, and Russia are in a race for the Bitcoin network.[ii] The large warehouses of Bitcoin miners are known and protected by competing superpowers. Any dreams of decentralization are hard to sell if these militaries took control of the warehouses.

that was fast

Bitcoin isn't even two decades old, and it's obvious to those willing to admit it that the entire mission of running a decentralized database set has largely failed to materialize. The entire network is centralized by a consortium of companies in the largest nation-states. The need for hardware and electricity compromised the solution.

Was this a market failure? Or just something specific to Bitcoin?

The problem with the market for new Bitcoins is that it auctions Bitcoin for random numbers with no control for the distribution of the winners. Since a wide number of winners was one of the de facto objectives (e.g., decentralization), the system struggled to meet its goals. In the same way, our merit-go-round didn't control for the quality of the input; Bitcoin not controlling for decentralization has led to issues.

Proof-of-work, like almost every market, is subject to downstream effects of economies of scale. There are real-world constraints that determine the winners and losers in these games. In more specific terms for Bitcoin, the system is dominated by those with an advantage in energy costs, chip manufacturing, and security.

We can see this trend play out in countless industries. The game is free-market competition along price and innovation until it's not.

Regulations come in and determine the winners and losers. The game turns into one of lobbying.

Firms end up driving costs through the roof on competitive marketing spend and the price increases for consumers. Marketing budgets rather than price or quality determine who succeeds.

Creative financing and investments cause startup firms to compete on securing investments rather than product and pricing. Sustainable long-term businesses are undercut by venture capital looking to flip equity.

The inevitable lesson of this system is that markets almost always compete along lines you don't expect. The market interacts with countless areas of the economy, and it's hard to tell which will be the constraint and if the resulting competition will be beneficial. Bitcoin learned that if the input is concentrated, the resulting market is too. Whether it's hardware or electricity, both have serious real-world complications.

why didn't they update?

Reading the story of mining centralization from the vantage point of Bitcoin as a company might be confusing. "Why not just update the algorithm and fix the problem of centralization and energy usage?"

To understand why Bitcoin couldn't change, it's important to remember that they wanted to be neutral. The level of this neutrality, however, was questioned. In fact, this was historically the issue of greatest contention within the community.

As Bitcoin began to grow, it had goals to stay decentralized, but it also had the goal of being accepted for payments. The early users and developers wanted it to function as a currency. Unfortunately, it became obvious early on that payments would be hard. The database was limited in size, and even by 2014, transactions were getting expensive. Everyone knew it was only going to get worse as adoption increased.

Staying decentralized, while also fast and cheap, is an open computer science problem, and although the technology did improve in the first decade of Bitcoin's life, only very recently did we get the ability to do finance anywhere near the scale needed for mass adoption. So, the option for Bitcoin in the mid-2010s was pretty clear: stay the same and stagnate at a technological level or keep experimenting and add in the challenges of dealing with competing interests in figuring out what to change. The argument for the latter was that we could change the system often, namely by increasing the throughput as the technology progressed. That way, the system could keep acting as cash and start onboarding new users.

On that side, you had several prominent developers and community members who wanted to increase the throughput of Bitcoin and continue iterating with the system to get payments right.

But some didn't see it this way.

On the other side were community members who believed that the politics of changing the system weren't worth the squeeze. They saw risks in giving too much control to software developers, citing that the government could force them to add blacklists or backdoors. Changing the system, in their view, threatened the "censorship-resistant" nature of the chain itself.

Additionally, by increasing throughput, there was a further complication in that it would usually increase the hardware requirements for running and storing the chain. This would make it even harder for the average person to get involved as your machine would now need to store and maintain a bigger database. According to the community, the only way to provide the payment aspect of Bitcoin would be to use other networks on top of Bitcoin (see the Lightning Network[iii]). To them, and a growing online contingent, the main utility of Bitcoin was not that it was a payment system but rather that it was an immutable asset that could be plugged into other payment systems.

The unchanging nature of the system was the value proposition.

The idea of only 21 million Bitcoins was a number forever locked. By having no risk of ever changing, it was a hardened and predictable asset – not digital money, but "digital gold." The real problem with the dollar, in their eyes, was that the Fed could print more. If cryptocurrencies were to be a foundational financial asset better than the dollar, they needed predictability. Protection against theft, censorship, and arbitrary rule changes should be the main focus, not actually being a currency. The system needed to be as "forever" as gold itself. For this to happen, the view was that the code should be hardened; it should never change again.

As with most things, the decision to never change, or "ossify," the chain benefited some companies more than others. In the Bitcoin debate, the firms seeking to control the payment channels and mining networks stood to profit if the base layer stayed the same, and so that's what they pushed for.

After years of heated debate, the issue was settled; the community did not update the code but rather shifted the narrative of what the code should do. No longer was it for peer-to-peer payments, but instead Bitcoin was an asset that would serve as a censorship-resistant store of value.

This shift was seen as a bait and switch to many in the community.

To the early developers and enthusiasts, Bitcoin was still an early technology that should experiment and adapt to fulfill the dream of what it originally sought to be. It seemed to them like this path was being put on the back burner for the short-term goal of making the asset more valuable.

The promise of digital gold was sold to the holders, and it succeeded. To half the community, it wasn't an attack in pushing through a bad upgrade but rather in not passing one.

escaping the system

It's easy to see both the reasons for Bitcoin's design and also why it failed to work out. The technical achievement of a decentralized ledger stands alone, but the implementation, in practice, could not maintain the principles it claimed to embody. Even if it is accepted that Bitcoin is supposed to be a store of value, it's hard to see that it's living up to its promise of decentralization.

Markets need hand-holding.

They can be set up with noble causes and well-thought-out designs, but the system will get what it incentivizes, no more. Just as the reward for weight led to waterlogged meat and private stores of fruit, proof-of-work has led to energy waste on the guessing of random numbers from a handful of state-adjacent corporations.

Even digital systems can't isolate themselves. The only reason the situation isn't worse is that the narrative around censorship resistance is enough to scare off potential monopolies. Eventually, even this dogmatic religiosity to decentralization will cede, and the community will need to come to terms with who actually controls their network.

~

The crypto meetups in the city were notably more distinguished than the hackathons. The corporate boardroom draped in designer office furniture was definitely a scene more familiar to Bennet, but he questioned the purpose of the event.

A boutique law office that normally charges thousands of dollars an hour didn't seem like the ideal place to find cypherpunk regulatory counsel, but according to themselves, they had apparently earned the title.

The room began to bustle as a few dozen people stood up from their seats. Bennet made a beeline for the speaker.

"Excuse me, um, thanks for the presentation," said Bennet. "Do you mind if I ask a few more specific questions?"

The lawyer continued to walk to the back of the room, motioning with his head for Bennet to follow.

"Shoot," he said as Bennet trailed behind.

"So, you said that the token has to be a utility token for it to not be a security? Does this mean that if it's being used for gas, like as an L1, you're good?"

The lawyer grabbed a Peroni off the table serving as the open bar and took a sip before answering.

"Listen, there's always nuance to this stuff. And how you raise funds is a separate question, but the laws are pretty clear here. If you just make it useful for something, it's likely a commodity. Kind of like gold, it's an investment, but it's not a security. You just can't promise a return."

His eyes scanned the room. He wasn't even attempting to look at Bennet.

Probably only a decade older, he wore an expensive suit without a tie and smelled like cologne. His presentation was on a recent SEC report relating to the industry. According to him, crypto regulations were clear to those paying attention.

Bennet wasn't convinced. To him, it seemed like selling a token that he could make in ten minutes was a security, but what did he know? According to the lawyer, as long as you could use it for something (even a chain that Bennet just created), everything would be fine.

Considering Bennet benefited from this answer, he didn't want to push too hard.

"One more question," said Bennet. "Can I ask what the penalty for security violations is?"

"Listen, if you've raised capital or want to work out an equity deal, we can set you up with some counsel to help design something that's compliant. But yeah, you know, as long as no fraud's involved, it's a fine. Jail time doesn't happen for these kinds of things."

"So, if I raise millions in a token sale to pay myself a big salary, the downside is that the company could be shut down from a fine?"

"Ha," he said with a pause. "You said it kid, not me."

3

CREDIBLY NEUTRAL DISTRIBUTIONS

"The Times 03/Jan/2009 Chancellor on brink of second bailout for banks."

- embedded message in the first Bitcoin block

One of my favorite characters in all of literature is Milo Minderbinder. The mercenary businessman in Joseph Heller's *Catch-22,* he represents capitalism and the economy. He's driven, hilariously and almost psychotically, toward doing anything and everything necessary to make money for his company.

Taking place during WWII, the company he represents is M&M Enterprises, or as he refers to it, "the syndicate." It's the global company that is entangled with everything. From food to war to production of all goods. It's inseparable from the economy itself.

Milo runs it, but we all own it. Americans have shares, British have shares, and even Germans have shares. As a result, even above the lives of our own soldiers, the subconscious drive of everyone is to grow the syndicate.

A tense moment in the book occurs when Milo is

confronted by the main character, an American soldier named Yossarian. Milo and his syndicate were involved in the death of Yossarian's friend, and Yossarian wants answers.

Milo, clearly frustrated, explains there was nothing he could do about it.

He ran the numbers, and this was the best option for everyone. He told Yossarian that the syndicate was contracted by the Americans to bomb a German bridge. The syndicate was also hired by the Germans to protect the bridge. Since he knew how to get around German anti-aircraft (he placed them) and he also knew when the attack would take place (since he planned it), the real decision was just how to keep everyone happy. Since the syndicate got an extra thousand dollars for each plane shot down, the planes were attacked, and the syndicate made more.

Everyone won because the syndicate won.

As ridiculous as it may sound, the most revealing part of the book is when he talks about how he wishes he could do something about it. Unfortunately, he explained, the Germans were also members of the syndicate.

"Don't you understand that I have to respect the sanctity of my contract with Germany?"

He wanted to save his fellow Americans' lives but was bound by his moral duty to increase the profit of the syndicate. If the contract insisted he increase the value of the company, the lives of Germans and Americans came secondary to the integrity of the contract. Being so entangled and entrenched with the syndicate, it was hard to argue for the lawlessness implied if the contract failed. It seemed that neither side had an option.

Everyone wanted to stand up for their dead friend, but no one could work through the details. Adherence to the system was so ingrained that it made bombing your friends seem like a valid option.

It all came down to numbers.

"Can't you see it from my point of view?" Milo repeated.

~

Bitcoin and early blockchains use proof-of-work, but Ethereum and almost all modern systems use proof-of-stake (PoS).[i]

The technological feat of Bitcoin was impressive, but proof-of-work (PoW) left quite a bit to be desired. Even besides the obvious environmental concerns, people wanted more out of a blockchain, and Bitcoin was clearly not up for new ideas.

Enter proof-of-stake.

Whereas PoW is secured by sunk hardware costs and electricity, PoS allows you to escrow the system's native cryptocurrency itself as the economic collateral you lose if you misbehave. In other words, PoS requires participants to lock, or "stake," a large amount of the native cryptocurrency (e.g., ETH on Ethereum) in order to become not a miner but a "validator." Validators then take turns choosing the transactions that update the database while also checking off each other's updates, specifically that no rules were broken. For Ethereum, this happens every 12 seconds, and if more than two-thirds of validators sign off on an update, the chain moves along. If someone breaks a rule (like trying to spend more money than you have), the other validators "slash," or delete, his staked balance.

PoS has the benefit that validators are not forced to waste money on electricity or expensive hardware to participate in the chain's validation. The only limiting factor is the ability to purchase the initial stake deposit, a very nice feature for young chains, as the cost of their tokens is relatively free.

Both designs are different versions of markets. They are rules for how participants can acquire newly minted tokens. The intended externality is the same (a functioning and secure database), but the nuances in how the next block creator is selected make a world of difference.

If you think of the market structure of Bitcoin mining as an auction at every block for the right to update the chain, for PoS, it's more of a lottery where your entries for each drawing are the number of tokens you have. For Bitcoin, the money spent on the auction doesn't go anywhere (solely to wasted electricity), but for PoS, you enter the contest through your stake, and as long as you hold it, you're entered in the next round.

A bonus for these systems is that the purchase of the stake should help to increase the price of the asset. In a world where there are no new entrants to the system and the users are the validators, there is theoretically no waste at all, just participants updating the chain cooperatively rather than competitively.

building different things

The vision for PoS was largely the same as Bitcoin's from the outset. These systems wanted the same decentralization, censorship resistance, and neutrality, only faster and more efficient. The bigger change that came with the launch of Ethereum was the ability for crypto to be used for more than just money.

For Bitcoin, the focus was on how to create a currency that would function even under direct assault from a real-world government. With Ethereum, there was still the dream of decentralized control, but now it expanded from just money to programming as a whole. A "world computer" that no one controlled and anyone could access. Whereas Bitcoin wanted to protect your wallet from inflation, Ethereum wanted to add functionality to money, digital assets, and more.

If Bitcoin is a simple ledger, Ethereum is a database where each entry can have rules attached to it. These "smart contracts" are glorified escrow accounts that give users the ability to add rules to their money. Now, instead of just being able to send money, you could theoretically recreate all of finance.

Ethereum started out as PoW but switched to PoS in 2022, with several other chains launching the mechanism first (Cosmos, EOS, Tezos, and many more). By the time Ethereum did switch from PoW, the cultures of the Bitcoin community and the "altcoin" communities (anything not Bitcoin) were starting to diverge. Whereas the Bitcoiners decided that they would never again upgrade the system for improvements, the other chains took a different approach.

These newer systems looked more like tech startups than Bitcoin. In addition to decentralization, they offered the narratives of opt-in and transparent rather than just immutable. To these founders, they recognized that to actually get usage on a product other than a digital rock, they would need to upgrade the software as the technology improved, and that's exactly what they did. Despite looking like a loosely organized corporation, there was still a near-religious dedication to the belief of decentralization, neutrality, and censorship resistance. Although the code of Ethereum could undergo updates, once something was validated on-chain, it would never be reverted. For Ethereum, the mantra became "code is law."

Tied directly in with this belief was the idea of markets everywhere. If the system were going to truly be censorship resistant, it would mean that the developers, founders, and even other users should have no say into how the system is used. From the selection process of the validators to the right to inclusion in one of the database updates, each piece was an auction.

Neutrality in these systems became synonymous with the privileged position of financial markets over all other forms of discretion. It meant zero say as to the content or purpose of the transactions, only a dedication to the financial gain to be made from a given action. If users wanted to buy drugs or expose state-level secrets, everything was game. If an attacker could somehow gain control of all of the coins, voting power, or influence on the system, this was all fair play. All distributional choices were handled by markets, not principles or personal judgement.

the dao story

About a year after Ethereum launched, a few developers wrote a smart contract that would act as an investment fund. The system would hold deposits of ETH, and then participants could vote on how to invest it. Your vote share was represented by the amount you had deposited.

It was relatively simple, but this was revolutionary. They created a democratic, completely disintermediated method for investing, and it represented the first of many experiments in decentralized governance. To the surprise of some (even the developers themselves), it attracted an enormous amount of capital. Within a month, over 11,000 participants collectively spent $150 million worth of ETH buying shares in the DAO. This represented around 15% of all ETH at the time.

It was exciting, but it was short-lived.

Like many things in crypto and new software in general, there was a bug. Someone found a loophole in the smart contract code and managed to steal it all.

Luckily for those participants, though, there was a pause on withdrawals, so the hacker stole the funds, but he couldn't get out of the system for 28 days. This is when the discussion happened: should Ethereum "fork," reverting the hack and returning stolen funds?

On the pro-fork side, you had people saying that we can do this and that we're irresponsible not to.

"It's just a database after all. Everyone holds a copy, and we should just update the values to remove the hacker's balance."

"Social consensus is real, and we should lean into it."

"The Ethereum network might not survive such a large portion of the entire supply being in the hands of a malicious actor. He could exert influence over the network, and his ownership might even limit adoption if the system is seen as benefiting thieves."

On the anti-fork side was the philosophical crowd.

"This would set a dangerous precedent. How are we supposed to stay neutral if we're making these decisions constantly?"

"Code is law; you should analyze and audit the code before putting your money in."

"Ethereum would prove that it's centralized by doing this."

It was a stressful week. Developers created a fork and then let the community decide. If users wanted to opt in to the fork, they could run the new code, changing things. If they wanted to stay on the "pure" code, they were also welcome to do that.

Long story short, the majority of users went with it.

The lion's share of Ethereum decided to revert the hack and save the victims. Of course it wasn't uncontested. To this day, the original branch of the code lives on in Ethereum Classic, a lesser-used network where the attacker still has his funds.

It was a success, but even among proponents of the fork, they talk about it in such a way that indicates, "It was necessary, but never again." Everyone in Ethereum claims adamantly that they would not and could not ever pull something like that off again. The developers and founders agree that the last thing any network needs is the complexity of choosing which systems to save and which are left to their own demise.

the continued story

Ethereum and other programmable cryptocurrencies were judged harshly early on. After debacles like the DAO, they wanted the respect that came from being decentralized and censorship resistant like Bitcoin; they also wanted to create a more usable system. Even despite the problems they had as early-stage startups, they saw the issues with the centralization of Bitcoin miners and the narrative capture that led to premature immutability.

As part of this, Ethereum and other chains worked diligently to instill the idea that their systems could be used by anybody, for anything. Every piece, from the users to the validators, would need to be decentralized and incapable of being stopped by the founders, companies, or even governments that had an interest in the chain.

It was this neutrality that characterized programmable chains.

Anyone should be able to launch a system like the DAO, an exchange, or a token, and the system should support it. There would be no choice of user or use case, but rather a support of whoever paid the most to post their updates. No favoring of incumbents, no picking of winners, and no bailouts.

As they realized with the DAO, though, it was difficult. How do you prevent a community from initiating a fork? Religious adherence?

It was all social coordination.

True censorship resistance meant that no one could ever fork. If the transaction was deemed valid, it would never be reverted. Ethereum understood this, but unfortunately neutrality stands in opposition to this goal.

fork choice and stablecoins

The main issue for programmable chains is that of fork choice.

Fork choice is the ability to determine when to upgrade or change a blockchain. This is the crucial piece for security in blockchains, and it directly affects how centralized or censorship-resistant the chain is.

To understand, let's explain blockchain security a little more:

Under most PoS rules, each validator takes turns proposing transactions, and then everyone else either signs them as valid or invalid. If greater than two-thirds of other validators sign it as valid, the chain is "finalized," and a block is made.

It looks from a quick glance that our simplistic PoS system is secure up to two-thirds of the validator set being compromised. If someone bought or bribed two-thirds of validators, they could break the system and steal all of the tokens or stop me from sending tokens, right?

Not quite.

Whether it's PoS or PoW, the system always has the option to fork; or in layman's terms, kick out the attackers and move on.

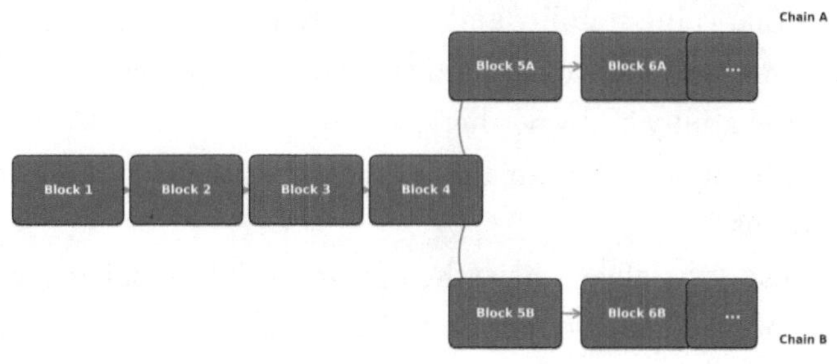

Blockchain Fork: One Chain Splits into Two. Author's own figure.

This ability is usually reserved for things like software bugs or system upgrades, but the important addition is for attacks. If the community as a whole decides to update the rules of the database to fix things (or revert a hack), they can. As Ethereum realized shortly after the DAO hack, no one has to keep using a captured system.

Ultimately, these are social systems that rely on social acceptance for their validity. Like a board game, if 5 of the 6 players get up and start a different game where the rules are changed, which game is the official one? Even if the lone player at the original table can claim victory in some technical sense, the society has moved on.

As with any traditional governance system, all blockchains have a society or "social layer" to fall back on, and this is who actually controls the database. If enough people agree to change the rules, then that's the truth. If we all decide to not use the current Bitcoin database and change to one without Trump having a balance, then the official "Bitcoin" changes.

This ability exists for non-digital systems too.

If we all agree to use cigarettes as the new currency, dollars would lose their value. If we all agree that the constitution is no longer valid, it ceases to be. The validity is social. Of course, coordinating such a move where all of us update our databases at one time is insanely difficult. For distributed and anonymous systems that value stability and neutrality, making values-based changes to the chain can be hard to garner acceptance.

But the ability is always there.

This ability to revert an attack is what ultimately secures all blockchains.[ii]

Honest users always have the option to simply delete a bad actor's tokens, remove an undesired block, or change the system and continue as if it never happened.

Why would someone attack a system if they know that the system will pretend the attack never happened? The only reason would be the attacker hopes they can exit before the system reverts, or they steal an amount too small for the system to coordinate a reversion.

The entire market system (PoS or PoW) is simply put in place to make an attack less likely in the short term. It makes it so communities don't have to, and hopefully never have to, coordinate an update. If the system can hum along without delays (e.g., coordinating a new database), it makes itself more useful to everyone.

In many crypto communities, the term "fork" is often seen as a dirty word. A system changing its rules or reverting a transaction that was executed is often seen as the antithesis of blockchain protocols. In reality, the idea of community-discovered truth is a feature, if not the feature, of these systems.

The problem comes into play when one party can control which of those "forks" is the official one.

This is fork choice.

picking the winner

If one group has all of the "value" on the chain, they can control the validity of a given fork. As an extreme example, if 95% of the users hold ETH on one centralized exchange and that exchange gets hacked, it doesn't matter if Ethereum the protocol is censorship resistant. It's likely the system will fork to save the users because they represent the social layer.

They control the fork choice because they can determine if the system reverts.

This risk is easy to see with banks and exchanges. Whether dollars or crypto, if they hold enough assets, they represent a forking risk. If they get hacked or lose user assets, the system needs to save them. For programmable chains, the biggest risk is stablecoins.

For most crypto assets (i.e., ETH, TRB, Bitcoin), if there's a fork, the coins can have value on both chains. The two systems can literally create two currencies (e.g., Bitcoin on the original chain and Bitcoin on the forked chain). Stablecoins and real-world assets are different. If a company mints you a coin that represents a dollar in a bank account, that company only has one dollar in the bank. As a result, they get to decide which fork of the chain can redeem the dollar.

And this is the problem currently for Ethereum and most other chains. If the stablecoin issuers pick one fork, all of the applications using that stablecoin must follow the fork chosen by the issuer.

The reason blockchains care is that the controller of the fork choice can often be coerced. If the US government went to the stablecoin issuer and said, "Fork the chain to remove this bad actor," would they do it? Considering the US government can literally take the dollars in their bank account, they would have to comply.

The size of certain applications is handing the control of these chains to US corporations.

This sounds extreme; however, if stablecoins were around during the original DAO hack, it's extremely likely that they would have been the ones who determined the correct fork. If the major stablecoins lost money as a result of the hack, it's unlikely that any philosophical argument could have been made to sway rational users against their interests.

Rational parties might wonder why Ethereum lets these external parties get so big. The issue is the adherence to a naive neutrality. By being open toward any system building on it, they are even open to systems threatening to control it. It sounds blasphemous, but Ethereum and other programmable chains can't let stablecoins, exchanges, or any application get too much control. They become too big to fail, and the system's main value proposition is rendered meaningless.

In the traditional government realm, this idea of fork choice is usually reserved for the military or some combination of large corporations. Even if all of society besides the military wants to oust a president, the people with guns might stand in the way. In the same manner, if they want to oust the leader, they probably have a decent chance.

It's a hard problem, and the even harder part of the question is how to prevent it. For our blockchain systems specifically, if they want to be credibly neutral, how can they be neutral towards things that threaten their own neutrality? How quickly should they remove the centralizing forces? How quickly can they?

These exchanges and stablecoins have become the Milo Minderbinders of the crypto world. They touch every blockchain system as well as legacy financial system. They exist and function on every chain and in every application. The chains can't fight back or stand up to the system because it would hurt their own bottom line. Even if they could get over that fact, the very action would violate their own principles of neutrality.

They've created their own catch-22.

the inevitable oligarch

Moral hazard and network effects exist as soon as parties get too big.

Once the idea is rumored that a chain would rather fork than let a subset of validators be slashed, unwelcome incentives arise. If Coinbase, a popular US exchange with millions of customers, owns 25% of the validators, would we really slash them if they misbehaved? It's a tough question. But what about 75%? Almost certainly.

The question then becomes what they will do differently when they have this implicit guarantee.

Everyone knows of the moral hazard present in the banking

sector. Since the FDIC will bail out deposits, Fannie and Freddie will backstop mortgages, and Sallie Mae will buy all the student loans, there's an inherent disconnect between those issuing the credit and those taking on the risk of default. As a result, you have private issuance of liabilities and a government guarantee of solvency. This is clearly unstable, and the 2008 financial crisis validated this theory. Risk was mispriced, and incentives were misaligned. Investors were bailed out as the banks were deemed "too big to fail."

Unfortunately, fixes were not adopted by the government. Beyond social embarrassment, there were few repercussions for the executives, and therefore zero reason for them, or new entrants into the system, to change. In fact, by bailing out the banks, the implicit guarantee was solidified and the problem exacerbated.

Crypto knows all about moral hazard.

The bank bailouts were one of the cited reasons for the creation of Bitcoin. But too-big-to-fail doesn't care about reasons or ideals. Even if a chain avoids articulating any explicit guarantee for concentrated validators or users, the silence may prove less protective than its advocates imagine. History is full of examples showing systemic threats forcing interventions that no one initially planned to make. Of course, the entire ecosystem is unwavering against giving said bailouts; however, desperate times lead to unplanned outcomes.

No politician alive would have said the banks were too big to fail, but it happened.

Justifications are created to prevent short-term pain. Crypto is in many ways backing itself into the same moral hazard scenarios it had hoped to escape. The truth is that as centralization increases, the odds of a bailout to save that party increase. As those odds go up, their appetite for risk increases, and the entire system is threatened.

Whether it's the centralized exchanges, the stablecoin operators, or even the founding team, they almost always start out as well-meaning community members who want to make the chain safer or more usable. But the success of these centralized entities is in and of itself the issue. They grow too big, and they end up controlling the chain.

markets as silos

Somehow all of crypto has been convinced that using a tokenized representation of a dollar in a bank account was the goal. They pay the most to have their transactions on-chain, so we have to allow them.

We're neutral, and the market dictates that it all comes down to numbers.

"Can't you see it from my point of view?" said the founder.

If the idea of what the chain should be used for is left to a market, there's no ability to check that the ending distribution of control is sufficiently decentralized.

Unfortunately, Ethereum and the crypto community left the original DAO with the wrong lesson. The idea that code is law and that you shouldn't fork to save users is important from a moral hazard standpoint, but preventing applications from owning the system should be priority number one. If a system's value is tied to one user or application, the system will become dependent on that user.

Distributional issues are the same as capture.

As seen with the refusal to update the Bitcoin codebase, users or validators can grow too large and capture the system by their sheer stake and control of it. Neutrality towards distribution destroys the claim of neutrality.

Even more difficult for these systems to admit is that there's no codified solution. Markets and tweaking the code cannot work.

It's impossible to measure or control for, as the influence

exists outside of the system itself. If someone launches a token on your chain connected to a real-world asset or legal contract, these might be completely private to the validators or other users of the system. There's no way for the community to know it's being coopted or threatened by the new user. The founders and developers who start the project become the minority as they allow the system to be sold to these new users.

If your system is open for purchase by the highest bidder, you don't get to say what's put on it. The distribution is not controlled for, and giant users or owners of capital take control and choose the narrative.

It's a two-edged sword.

The Bitcoin community discovered during their own update debate that even if the founders and developers want to change the code, corporate interests and external parties can muddy the dialogue.

"Maybe giving these experts discretion over the codebase is a good thing."

"Maybe their ability to fork is the right move because they've only done it to bad people."

"This setup allows institutional adoption of crypto, which we've always wanted."

These justifications have legitimacy, yet they often represent a fundamental departure from the system's founding principles. Intermediaries and parties benefiting from increased use of their systems will find ways to make it seem like a good idea. They pollute both the narrative and discussion around the truth or even the historical goals of the system. The community will feel like it was a conscious choice made to increase the value of the asset.

These protocols now face a choice between their theological commitment to neutrality versus any coherent articulation of an objective.

A system that doesn't allow a validator to censor could be

seen as limiting his freedom of expression. A chain that doesn't allow one user to control its fork choice is restricting their usage of an open protocol.

It's a paradox.

Can a system claiming neutrality remain neutral toward forces that undermine that neutrality? If you desire no one to hold an opinion, are you enforcing your own?

~

[team AMA on Dojima telegram group]

the_Bennet: Hey Dojima community, thanks for joining the AMA! To just give some announcements before we get started, we gotta talk about the upcoming hard fork. I know most of you saw the proposal to upgrade the system. If you didn't, link's here. The TLDR is that we're moving to PoS and all further upgrades will be managed by our new governance structure. We had an amazing launch so far this year, but we need to keep moving. Now that it's live, we can see the issues, and we all know we can do better. The main points are the inflation and the centralization. The system has a very high inflation rate that many of you are concerned about. It was needed for distribution early on, but we agree it needs to come down now. The other piece is the PoW structure and the centralization. Since the FPGAs came out, the open-source pool and even the team can barely compete. We've done the math and shown it around and PoS will allow us to have more security as well as the ability to direct some inflationary rewards to a grants program that we can use to get some more users to build on the platform. On a side note, too, just so I don't forget, I'll be giving a talk on the new structure next month in New York, so definitely come out if you're around. But I'll let Josh add anything else.

darkSith: Hi. The system upgrade is ready for review in the github. The system will undergo a hard fork to perform the upgrade. Be sure to check the discord for the exact block height when we announce it. We'll be going live on testnet starting in 2 weeks. We have a few oracle and bridge partners we'll be using at launch also.

the_Bennet: Thanks Josh. I think that's it, so want to take over salad?

saladToes: gm all. I'm just going to post the top ten questions that were voted on by you guys. The open AMAs just lead to trash "wen moon?" comments or straight up harassment, so leaving the chat off for now. Will open it up later. Also, please remember, as always, there is no reward or airdrop for participation in the AMA.

But the first question (I added quite a few together for this one): "When do the team's tokens vest and what protections are there for the holders?"

the_Bennet: Yeah, so we're vested at the 6-month mark which is next month. I know underneath the question is some fear and a question of commitment. A lot of times teams dump the tokens and leave the project but that's not us. Just rest assured that me and the team remain as committed as ever to the project. We launched this in the bear market because we believe in what we're building. We're all doxed and we're putting our reputations on the line for this. I know for myself, I'm in it to drive this thing to mainstream adoption. The power of open systems and a world where finance is transparent and fair is coming and we're super excited to be a part of it.

saladToes: The next one from our pool operator shep: "What are the plans for the current PoW miners?"

the_Bennet: Great question. As we transition, you'll have to transition to actually buying the tokens vs the hardware. We know this is a little bit of an ask, but we did originally target GPUs which are repurposable (or even profitable to sell), so the only people who will likely lose are the FPGA miners and producers, who to be honest, are a big reason for the upgrade. They should have plenty of tokens by now though.

saladToes: "When are we going to be immutable? All of these upgrades and forks just show how centralized we are. Did you guys give up on your ideals?"

the_Bennet: Ha, no not at all. We came into this space with belief in free markets and decentralization and we're still all about them. I'd say we're just moving with the space. We realized that PoW is centralizing unless (or even if) you're bitcoin and we want decentralization. You're more than welcome to run and try and convince everyone to stay on this fork, but we need to push the system to something that users are demanding. In addition to less emissions, we'll also be able to be much faster and utilize the amazing research done in the field. The new governance path too is inevitable. Everyone is moving to these new structures and it's clearly because these systems need a way to upgrade and take some input. The contracts we've made allow our stakers to vote on everything, and it's the future of building an entire government out of our ecosystem. It's really making the system bigger. Instead of just being a piece of code that works for finance, we can expand to encompass everything. We can be an entire nation online and it's exciting.

saladToes: This one's from me, "Who will grants go to, and how will that be decided?"

the_Bennet: Governance will!! We're gonna vote on it, so get ready! Really no comment past that lol.

4

PUBLIC GOODS FOR THE CHAIN

"Indeed, unless the number of individuals in a group is quite small, or unless there is coercion or some other special device to make individuals act in their common interest, rational, self-interested individuals will not act to achieve their common or group interests."

-Mancur Olson, The Logic of Collective Action

Jay is a cypherpunk who's been around the internet longer than most crypto founders have been alive. His dream, like the dream of many early internet developers, was that everyone would run servers in their house.

That community's vision for the internet was one based on peer-to-peer interactions. Everyone would store their own data and interact with the world on their terms, not those of some corporate intermediary. Jay tells stories of running his own email server and websites out of his basement. There weren't many people online back then, but they were doing it out of passion and hobby rather than any long-term financial freedom agenda.

Of course, being nostalgic for a time when you were a younger man is nothing novel, and it's hardly an indication that the situation was sustainable, but the vision was there. He and many others recount similar stories of the good old days and how things slowly changed in the early 2000s. The internet slowly began to arrange itself with centralized parties running the infrastructure, namely cloud providers and hosted infrastructure. Financially, it made more sense to have one person run everything. Companies could have faster webpages and store more data about their customers.

Jay saw crypto as a way to bring back the dream.

Everyone would run their own node, store their own data, and interact with others on a voluntary basis. Users would share only what they wanted to, front ends would be locally hosted, and everything from the banking rails to commerce would be done in a completely peer-to-peer fashion. This was the vision in both the early Bitcoin and Ethereum communities, and Jay jumped at the chance to help deliver on it.

"Run your own node," he would say to anyone who would listen. "The network needs people to care, and it needs you to verify the chain."

Jay understood that genuine decentralization required lots of node operators. Blockchains need individual participants maintaining and verifying their own copy of the database, or else users would just be picking different middlemen. To Jay and the early crypto developers, decentralization meant that anyone could run it and everyone was running it. It was a return to the peer-to-peer model because they felt that freedom could only come through personal ownership and operation. Any reliance on third parties was just setting yourself up for failure.

Jay knew this battle wouldn't be won easily. The early internet failed to keep people running machines at home. The user experience of just clicking a website and letting it store your data proved too tempting for most users to avoid, so early crypto developers tried something different.

In the early days of Ethereum, if someone wanted to hold funds or interact with the chain, they first had to download a piece of software called a node. These nodes would connect to the network, download the chain's database, and keep it up-to-date. The view of early developers was that users running nodes would access websites that had no central server; each person would operate the front-end on their own machine, connected to data from their own verified node. Users would hold their own funds and hopefully even contribute back to the open-source software that the systems relied on.

It worked, but this was only the beginning. As time went on, people realized this was difficult for new users. Running one of these systems wasn't trivial, and it took time. After a few years, crypto did pick up traction. Unfortunately, the new people entering the system weren't developers or even familiar with cypherpunk ethos, and they just wanted to use the systems. They were put off by the effort necessary to stay "pure."

The databases, for instance, took so much time to download that it would often be hours or days before a new user could send funds. This was a nightmare-level user experience, so people found ways around it. Seeing the business opportunity in the demand for an easier onboarding, some entrepreneurs began letting new users just interact through a centralized database. "Just trust me, connect to my database, and then use it as if you were running your own," they said.

This was the beginning of choosing user experience over the purpose of the system. These "hosted node providers" began to take off. At first, they acted as a backup for teams and users. Very quickly, however, it became the default. Wallets and websites would just build the hosted integration directly into the product. By the first signs of true adoption in the early 2020s, they represented almost the sole way new entrants and the vast majority of all users interact with the network. In fact, now most casual users don't even know there's a need for a node.

These private companies make money by making blockchains easier to use, but over time they compromise the network. It's the exact same problem that web2 experienced. Rather than Jay's vision of everyone owning their data, there are centralized intermediaries that gatekeep the system for the majority of users. There are even further downstream effects as these private companies begin to lobby the chains themselves. In the case of nodes, this means increasing computation and hardware requirements. If it takes a professional devOps team to properly use your system (run the database), then it quickly leads to only professionals doing it.

To state the problem more bluntly, in order to be decentralized, chains need people to run nodes, a sometimes-costly task. Since it's easier to skip and there are options to avoid it, the unincentivized task goes unperformed. Just as with the internet, consumer behavior led to an undesired outcome.

The lack of funding for needed infrastructure is a problem not at all specific to crypto. Most crypto protocols are less than a decade old, and even from the start, the lack of incentives to build certain features or run the system in a certain way is leaving gaping holes in these ecosystems.

the tragedy of the commoners

Public goods are the kinds of things that everyone benefits from but no one owns. In economics literature, they are opposed to "private" goods or things that are owned or used by a specific person or group. The classic examples of a private good might be some food, a watch, or even a phone, where one person owns it and it's for them. A public good, on the other hand, is something like roads, the court system, or a lake. No one really owns them, and everyone benefits from taking care of them, but it's not directly one person's responsibility to do so. The environment is one of the most explicit examples of a public good; everyone lives in it and wants it to be taken care of, but it's hard to make others pay for its maintenance even if we wanted to (e.g., how do you force a payment for clean air?).

Oftentimes these sustainability jobs are given to the government, but the general idea of public benefits with limited ways to enforce paying for them is a problem for all companies, markets, and economies.

"Fix the dilapidated roads."

"Provide paid maternity leave."

"Go clean up the trash in the ocean."

"Run nodes."

The examples of unfunded things that are universally agreed upon as beneficial are too numerous to list. If I pay for it or push for it to be adopted, we all benefit, but there's no way to exclude others or charge them for the work I put in. "Why is it fair that I build a public park, and you get to use it too?" This attitude prevents us from even starting construction, when in reality, we all want it done.

Part of the issue with public goods funding can be seen as more of a critique of the term "public." A "public good" sounds evenly distributed, but in reality, nothing is completely a public good or private good; there are levels to the degree to which a public good actually helps everyone versus just one specific group of people. There are certain public goods that I use more than you and vice versa. If I have children and you don't, you may benefit from having a good education system in our city, but not nearly as directly as I do. The same can be said of parks; maybe I use them more since I have kids or just live closer to where they're located.

Not all public goods are equal in their benefits, and there's an inherent tension that exists between something being good for all of us and the "she gets more than I do" mentality. As a result, we end up in a world where each individual (or small group) is pushing for their own "public good" from the shared coffers, and we all feel resentful when the other person gets more. At the same time, we all recognize that just privatizing the effort doesn't really solve anything. No one wants a park we have to pay to use, and for the broader environmental problems, even the concept of privatization is difficult to conceptualize.

And this is the crux of the problem: we're all competing, so a public good that I fund more of, or that benefits you more than me, isn't really in my interest to push for.

Even if you assume, like a pure libertarian, that everything is a private good, there are obvious examples where human cooperation is beneficial. Historically speaking, this ability to coordinate the funding of some greater good often decides which larger group of people will succeed. The classic example is that a country that can coordinate a wartime production effort is in a much better position than one that can't.

Corporations are a solution to this problem at a smaller level. Regardless of what you think about the employer/employee relationship, it's a general agreement to use a pool of resources and give each person a role so that a greater goal can be accomplished. Labor can be divided. Whereas one person can't build and operate a giant factory to reduce costs, the corporate model makes it possible.

Unfortunately, crypto protocols don't work like a corporation. Instead of sharing everything under one umbrella, most protocols try to structure themselves as a sort of immutable piece of infrastructure. They aim to be a system with no centralized operator or controller, the libertarian dream of only private goods with aligned incentives. A phrase you might hear is that "all participants are self-sovereign individuals cooperating solely for incentives." As always, it's a mix between private and public, and despite the claims made by some crypto enthusiasts, these ecosystems tend to suffer from similar market problems to the ones they purport to replace.

public goods in crypto

The often-unstated goal of blockchains can be crudely summarized as "increasing the value of our digital asset by selling blockspace (database entries) to the highest bidder." The narrative of "why" there's blockspace in the first place can change or even be unstated, but there's generally a go-to-market strategy of trying to get developers to build things on it (other tokens, payment systems, financial applications, other blockchains, etc.). The theory is that as the demand for these applications built on top grows, this will drive usage of the database and thus the price of the asset.

As with any platform, there are certain things needed to get developers, users, or even investors of the asset across the line to purchase this blockspace. Just as a traditional software company might need a website, some marketing, and a help desk, blockchains also need some basic infrastructure. For a traditional corporation, the company selling the thing will simply build the tools necessary. If you take AWS, for instance, Amazon maintains the user interface, does the marketing, writes the documentation, runs the data centers, and does literally everything else needed to keep the customers coming. People will then pay to use their servers (or "blockspace" for this example). There's a direct relationship between revenue from their product sales and the amount of money the company makes (more sales means more money for salaries, hardware, marketing, etc.).

For crypto it's different.

There's often a founding team, but they actually can't control it like a company. Mainly for regulatory reasons, but also because it throws the entire claim of "decentralization" and censorship resistance out the window if one party owns most of the assets and calls the shots. So, the founding teams only take a portion of the tokens.

In regard to funding certain goods, a critical issue is that the revenue paid to these systems (fees to make a transaction) is not given to the founding team or employees, but rather to validators (or sometimes to token holders).

In the AWS or cloud provider example, this would be a structure of cloud providers renting their hardware to the network. So, a user would come to the system, and the revenue would get passed to the independent operators of the datacenters. This could be an interesting system, but if 100% of the revenue is passed on to operators, issues begin to arise as to who will create the user interfaces, design the coordination tools, or perform countless other tasks for the network as a whole.

As chains have realized, there's a problem regarding incentives.

When it's no one's specific job to do a task, it becomes very important to have a community and not mercenaries. More specifically, not all validators are equal. Some are developers who build on the platform, some are die-hard believers like Jay who sell the product at meetups around the country, and others are just custodians or even speculators who will dump it the second it's unprofitable. The chains try to make close-knit "communities" or ecosystems out of their products to keep some religious attachment, but open systems that enable exit are hard to keep pure. Very quickly in the lifecycle of these chains comes the problem of public goods funding.

How do we pay for the base needs of the protocol?

How do you pay to build wallets, upgrade the software, host documentation, or even do marketing for developers to get them on our system?

As you might have guessed, the validators are all unrelated

entities. Unless you tax all of them, it's unfair if one of them bears the cost to build these tools. It's a classic public goods problem. The usual answer is that a "foundation" will get some tokens at the start of the chain, and it will be their job, but with no form of revenue, it's a temporary solution at best.

Two specific examples that show market failures in this context are block explorers and user infrastructure.

block explorers

Block explorers are a different kind of public good than nodes. Whereas you need lots of people to step up and run nodes, you really just need a few good developers to build and run an explorer. The problem, like the maintenance of the chain itself, is who funds the developers of these systems?

A block explorer is a website or piece of software that lets you view the blockchain and all of the data in it.

The database of a blockchain is terabytes of non-human-readable data that needs to be converted into something we can interact with and understand. Of course, it's nice to have your own node verify information, but being able to look at the information, even from your own node, is just as important as having it.

The challenge lies in establishing sustainable funding for this software.

For most chains, there's usually one explorer created by the founding team, but it's actually not that decentralized to have development, hosting, and maintenance of the website done by the team. If you add in node hosting (running the database for users, also often done by the team) the entire system looks a whole lot like a centralized company operating a database. Teams realized this, and many teams decided to leave certain pieces, especially websites that regulators might see, "to the market."

To fill the gap, private companies set themselves up to deliver public goods for purchase. These explorers represent problems for the chain because they're centralized entities, mainly funded by ads, that can censor or alter what parties see. A good explorer should run on your own computer or be verified by your own node, but now we're back to just trusting someone. The entire premise of decentralization is thrown out the window because the team can't figure out how to fund a website that lets users see the database.

user applications and rollups

The next public good example is similar to the block explorer in that you need to fund it but can't due to resource constraints or just ideological differences. For most blockchains, they require people to build on top of them. Other than Bitcoin, which mainly focuses on its "use" as a digital rock, blockchains want to be used. They want novel applications built on top of them. The theory is that if there's more usage, there are more people holding their token and using it to pay for database updates, and as a result, the asset goes up in value.

So, these chains run hackathons, sponsor educational content, invest through venture funds, and reach out to "builders." These are people starting companies or running decentralized applications that will use the underlying chain. Examples could be a prediction market, an exchange, or a new social network. The chain itself isn't interacting with the users; the companies running the application are.

Unfortunately, there are some base infrastructure and applications that need to be present to even get the system usable. Notably, things like bridges, oracles, exchanges, stablecoins, and lending protocols have all become applications that users expect on new chains.

To explain, imagine that chains are office buildings.

The business model for these chains is the hope that by renting out office space (database entries), profitable startups will come and create amazing businesses in their building. As each startup grew successful, they'd either buy more office space themselves or network effects would kick in, and other startups would come to join them. Eventually, the demand for their office space would lead to an enormous increase in the value of offices in that building.

The first issue for blockchains is they all need the same service providers to even start.

As you can imagine with our building, in order to be useful for even the first tenant, the building needs things like elevators, parking, air conditioning, and electricity. Rather than just internalizing these costs and including them in the rent, the building operator views them as separate tasks for different teams. So before even being able to service a single real user, they have to attract an elevator company, a parking service, an HVAC business, and even someone willing to sell electricity to tenants.

Of course, the same handful of "infrastructure" companies work for every building in the crypto ecosystem, but that's beside the point. Oftentimes, the chains spend enormous sums of money trying to attract these companies to build the base of their system. If they don't, it's often on the users themselves to go and procure these companies to come to the desired chain.

These products, which should be public goods available to all tenants, are instead given to the market to deliver.

What we're finding is that competition in these areas isn't developing. Only a handful of firms operate these necessary tools that every chain relies on. Decentralization is a hard claim when the bridge provider, oracle, exchanges, and RPCs are all operated as local monopolies by US corporations. Even worse for these chains is that the values of these service providers very rarely align with those of the chain.

As separate systems with different values and goals, it's no wonder they don't work as extensions of the base chain. If the provider is fine with scamming their customers or lying about security, everyone ends up looking bad, and the base chain often has no say in the end product (these chains are "neutral" as to use case). Even in a less extreme example, we've seen time and again that the blockchain will be "decentralized," but the bridges, oracles, and exchanges will be completely centralized. Users lose money due to security issues, chains freeze funds to protect corporate interests, and the entire idea of "no intermediaries" is an illusion that vanishes upon any scrutiny. Often it is their fault for misrepresenting their own abilities, but it's also a predictable outcome when these competing firms need to (or can) cut costs to make a profit.

At the end of the day, it's the same problem as the block explorer. You need someone to build X (be it a bridge or a website). Because the system can't figure out how to fund it or would be centralized if they did, they leave it open for anyone to do. Someone comes in and does it, but they have different values and often run it in a centralized fashion or take advantage of users. They're interacting with you for a return, not because they care about your goal.

Again, the neutrality compromises the value.

direct market failures

Funding public goods is hard.

If a community knows something needs to be done, it's best to own those hard decisions. Leaving both the existence and distribution of certain tasks up to no one is a recipe for poorly designed markets developing as a result. If there's demand, someone will figure out how to monetize it, but now the structure of the product is out of your control. If there are any values you hope to keep, it's unlikely they'll be voiced in the new design.

Some might argue that this is not directly a market failure but actually a governance failure. There's some input being a chain's revenue (usually from usage or investment), and then there's output in things that need funding, be it developers, running nodes, validators, creating applications, or even scaling your chain. The mismatch is a problem of allocation and distribution.

If the chain and its community internalize the cost and the decisions, it can be handled properly, but it's hard work. It's hard to come to an agreement, and it's even harder to do so in a fair and decentralized way. Governance and politics are messy. These problems are impossible to solve with code, and tweaking the parameters cannot help.

If you leave critical tasks up to no one, this is still a decision. In this case, the "market" is the absence of a decision. It represents the lapse in taking responsibility and ownership of a needed decision. The fact that there is a market is often only because the obvious party to fulfill the task refuses to do so or can't figure out how to align individuals accordingly.

If families refuse to care for children and instead kick them onto the street at age 15, this represents the creation of a market. Even if the free market can allocate housing and food efficiently, the entire existence of the market is a larger failure of the society.

A government halting all funding for roads or judges is not "creating a market for roads and judges." This doesn't just allocate resources differently; it produces an entirely different outcome from the one envisioned.

The truth is that incentives need to be properly aligned to pay for public goods. If people have money and no one can get things done, it's a problem of coordination. Open markets are a poor substitute for figuring out how to coordinate the incentives.

We know that private companies fill the void where public parties fail to step up. But there are downsides. Even besides the unprofitable, underserved populations, private companies now influence and exert control over your open system.

the real world and the end state

Jay still runs his node.

He even helps build an open-source block explorer. He's one of the few that do it without incentive, but he knows he's a dying breed. And yet, he paved the way again.

Crypto has already won in some sense. It's survived almost two decades, and Ethereum processes more value than Venmo. The technology proved itself, and the foundations set by cypherpunks still live on. The question now is how to scale them.

When needed services are left to the market, private companies do step in, but they have different goals. They need to generate a profit and will only serve the demand that will generate one. Founders and leading community members need to realize that this failure to fund public goods is not "neutral." Despite what you may hear, there's no way to get around the politics of fund distribution. Founders and community members alike need to own the process and work through the problem. Decisions need to be made, and values enforced. Which specific incentives, protections, and regulations are needed or tolerated is still an open question.

~

[MESSAGE POSTED TO DOJIMA DISCORD – MARCH 04, 2021, 03:54:11 PM]

Authored by masterqueef

Buckle up, just wanted to give a little rant here before I never come back.

This place is a slow rug.

They blow money on useless shit and never give a shit about their community.

Dojima went from standing up for something to bending the knee for every institution or cex that wants to pump our bags.

I kind of don't really know where to start or where it went wrong, but as most of you know I've been here for a damn while.

I'm mainly a lurker, but I do my best to keep us all on the same page and even act as a mod when the team takes off.

Now the price is down like a rock.

This is all of our fault, and we've all needed to fix it for some time.

But since you won't, I'm out.

To step back, why did we do it? Why are we all here? Why do we spend countless hours talking about this stuff, debugging code, and running software that will likely never pay us back?

It's obvious, FINANCIAL FREEDOM.

It's why I got in.

It's why Satoshi got in.

We want the ability to keep what we earn.

A fucking meritocracy where we can keep what we have away from some money-hungry government of incels who don't know when to leave us alone.

It's not hard.

Just give us a token and a platform, and we'll do that.

But apparently it is hard.

This shit's just a database.

It's code that requires no intermediaries and no trust.

We all download the same thing and run the same code and it's equal.

No more Federal Reserve or kings who control the money supply.

Satoshi solved the double spend problem and the problem of who gets special privileges by inventing mining and it's beautiful.

I can run it, you can run it, and we can all compete.

Dojima was meant to be a meritocracy.

We're not asking for handouts.

We made a chain that people can use for finance and this is the point.

But what the hell happened?

Where did we go wrong?

We all see it.

The team and the VCs turned into kings.

There are two types of people and you and me ain't in the right club.

We raised from VCs and gave them cheap tokens… not open or meritocratic.

The founders changed all the minting rules

once they saw the easy exit of dumping their tokens…not open.

They abandon the miners that helped make the system great… not immutable.

The token's down 90% since they launched and they do zero marketing when they switch to PoS.

Bennet's design works, but he's off getting laid at every conference the VCs put him at and I haven't seen the biz dev team doing anything useful since he hired them.

They might mean well, but someone clearly doesn't and whatever they do ain't working.

The few of us who were in the trenches mining this shit unprofitably were left high and dry.

PoS just gives control to a few custodians and the rest of us don't have any say.

They gave two grants to these hackathon winners that ghosted us as soon as the grant hit.

The users are dry and all we have is some gambling apps with some copy-pasta shit from other chains.

Yeah, our chain's cheap but users don't care.

It's cheap because no one uses it.

I don't even have a solution.

Just fix it and guys like me will come back.

Don't be such sellouts and actually build something.

I know there's regulatory issues in the US,

but we need a god damn user, not better infrastructure.

And we certainly don't need tradfi or this institutional adoption shit.

Screw anyone involved in this dumpster fire.

5

SCALING THE CHAIN

"If I'm ever wrong, I must make use of all my means and resources to bend and align reality according to my mistake so that it ceases to be a mistake."

– Hernan Diaz, Trust

One of my favorite concepts in economics is the contradiction of "one more lane."

It's the oft-proposed solution to traffic. To the average person, if you see a city that has a traffic problem, you quickly arrive at the solution: "add more lanes." It's a pervasive solution amongst suburban local governments across the nation... "Let's widen I-whatever."

But it never works out.

Decades of transportation research have consistently shown that expanding roadways leads to proportional increases in traffic volume. You can never get rid of traffic, and even worse, you completely change the footprint and feel of your town.[i]

This is the concept of induced demand. Simply put, an

increase in supply leads to an increase in consumption. In the case of roads, time after time it's seen negating any congestion relief efforts.

It's slightly backwards at first glance, but it makes sense. If you're willing to commute 20 minutes to your job, you live in the biggest house you can afford that gets you to your job on time. If the city builds more lanes and you can live further out and keep your commute under 20 minutes, you do so.

And so do other people.

If businesses know they can get deliveries quicker, they might keep less inventory or even no inventory at all. If employers know there are workers in the larger region that are able to travel downtown, they build bigger skyscrapers for offices and force them to commute. More workers mean more demand for lunch spots and happy hour bars. So that extra lane changes your whole city. It's not instantaneous, but eventually demand moves with supply, and the endgame of more lanes is a giant sprawling city with long commutes and a car-riddled downtown.

scaling blockchains

Architects of blockchains face similar challenges. They need to pick how to balance scalability and throughput with livability and usability. Also, like many urban planners, they often defer these infrastructure decisions to market forces. The larger problem in both situations is that there's rarely a vision for what kind of city they're trying to build.

Whether it's upgrades, the roadmap, or even the narrative, crypto protocols often have no north star that guides them but rather a fluid system that's roughly coordinated around the price of a given asset and a vague dedication to the censorship resistance and neutrality of the blockspace (i.e., entries in the database).

The systems will launch, get some mild traction, and then almost instantly be used for things completely unintended (e.g., crime or gambling). The legitimate use cases are priced out, and the developers are back at the whiteboard trying to figure out how to bring down costs.

They can't figure out why there are never enough lanes.

WHEN IT COMES to building blockchains, there are known tradeoffs. The most notorious is that of the scalability trilemma.[ii]

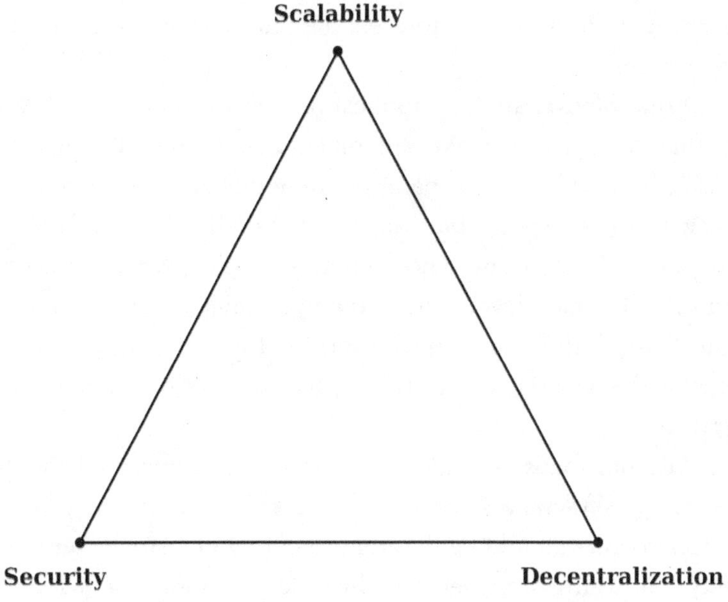

The Blockchain Scalability Trilemma. Author's own figure.

Stated simply, if you want to scale your chain (allow your chain to handle more transactions), you'll have to either give up security or decentralization. Barring some technical breakthrough, throughput comes at a cost.

If you have a network where every person on earth runs a computer program and verifies everything (perfect security and decentralization), this isn't very scalable. Just the time it takes to pass the information to every computer on earth is limiting.

As chains get bigger blocks and more data to send around, it takes longer. Storing this information requires bigger machines, and pushing it around the globe takes better network connections. Since bigger machines and better network connections cost more, fewer people can do this, and thus decentralization and security of the chain are reduced.

To put it even more simply, the bigger the data you're sending around or the more people that have to verify it, the slower it is.

What blockchain and application developers have realized is that you have to make different tradeoffs for different use cases. If you're trying to be a real-time global stock exchange, then you need speed, but maybe not WWIII-level censorship resistance. If you want to host a platform where whistleblowers can publish state-level secrets, then you might want the opposite. Same with finance, maybe you need sub-second timing, or maybe the next day is fine as long as you never have a reverted transaction.

Like our roads example, if you want to be a metropolis with booming skyscrapers and port access for the country, your infrastructure should be different than that of a small community that wants to preserve its historic row homes or working farms.

Knowing this, it's important to understand that the scalability of these systems is rarely about which team has the better developers or builds the best technology; it's much more about the tradeoffs each of the chains is willing to make and that they continue to make. If a chain selects one point in the tradeoff space initially, but the market signals that it wants more speed and less decentralization, the fundamental governance question remains unresolved. Who holds the authority to make these architectural tradeoffs going forward? And if it's not a person, what mechanisms should be used to make or enforce these decisions? We've shown with the Bitcoin example that technical stagnation leads to an inability to be useful for much besides speculation.

ethereum scaling

Ethereum began with the narrative of being a "world's computer." Anyone should be able to verify and execute any computational task in a censorship-resistant manner, with an even further guiding principle that they wanted home computers to be able to participate in the validation of the network.

But things quickly got crowded.

From ICOs and new token speculation to things as trivial as CryptoKitties, a collectible digital cat, it was felt as early as 2017 that their single chain was not nearly big enough. Ethereum developers always knew they needed to scale, but it had become even more pressing as transaction fees climbed into the tens, or in rare cases, hundreds of dollars.

In order to scale, there were two options on the table: scaling the base chain or just using the base chain to verify other chains, dubbed rollups. On the base chain scaling side, this was the trilemma. You had to make technical progress, as compromising security or decentralization was seen as off the table. On the latter side, it sounded like a cheat code. The strategy was seen not as a narrative shift but as a technical solution that allowed infinite scaling.

These rollups, or "layer 2s," would be new blockchains that

desired more decentralization or security. By using this new method, Ethereum could give other chains the same properties of decentralization that Ethereum had. The user chains would just need to pay for this property by sending some transactions to the base chain (e.g., Ethereum). By basically posting summaries of their new datasets on the more secure and decentralized base-chain dataset, end users could utilize the new chain with the same trust assumptions as they would the original.

There were good reasons to go down either path, but ultimately, circa 2020, they went full steam ahead with the rollup roadmap.

Fast forward six years, and it's not really working out. Those other chains did show up, but they didn't follow Ethereum's values. The interoperability and cooperation never happened. They paid lip service to them, but it wasn't the same. The worst part was that now the entire conversation became polluted. Pushback from other chains came on any scaling decision. Just as with any market open for purchase, companies built business models around the current structure, and now it's difficult to escape.

chains as pizza shops

To explain how this happened, let's imagine a blockchain as a pizza shop.

This shop sells delicious, high-quality pies. But they're expensive, and there's only so much room in the little shop. So, the owners come up with a plan to scale the shop. "Let's sell the ingredients, and we'll let other people make and sell the pizzas. Then we can eventually have one website where we route orders to your nearest shop, and customers won't even care who's making their pizza. We can even have joint marketing and rewards programs across the board for every shop that sells our pizza."

"The satellite shops will scale our shop."

The idea is popular. Lots of new pizza shops open up with plans to buy the ingredients.

But then after a few of them open, something happens.

Joe's pizza shop, owned by a longtime friend, starts selling their own pies too. They still buy some ingredients from the original shop, but they add a few of their own and even sell some items with completely different ingredients. They also just sell on their own website and even take most of the profit and credit for making the pizza.

Pretty soon all of the shops are doing this.

They're not really interoperating or sharing customers because they don't really want to send customers to other shops. They also rarely hold themselves to the same high-quality standard that the original shop had. That said, they still like the initial customer boost they get from saying they're one of the satellite shops (a boost getting smaller with each new shop).

This is what has happened with scaling crypto.

Constraints limited the throughput of the base chain, so they scaled through satellites, or rollups.

Now, like the pizza shop, the options are limited.

They could just work on making a bigger shop. They could even run lots of shops in other towns, all coordinating around the same corporate structure. But it almost feels too late for that as they're getting pushback from the satellite shops. They're big businesses now, and they don't want to compete with the founding shop.

Many of those who've been eating pizza at the original shop for years say, "Screw the satellite shops." Sure, scaling was part of it, but these customers just liked the high quality and welcoming environment at the shop. The stockholders take the opposite view. They actually want to start selling the ingredients to grocery stores and corporate pizza shops.

It's hard to say what the right choice is.

do you want to live here?

Roughly since the inception of crypto, chains began to compete on transactions per second and cost per transaction. They would calculate how many transfers they could do in a second and then tout with undeniable certainty that their number was the best and fastest. It was too easy not to. Of all the ambiguity in the terms, this one metric is easy to compare and quantify. It ignores the infinite nuance within the tradeoff space that exists beneath these top-level aggregations, but the markets loved it.

They rewarded the number.

The faster you could be, the more your token could be worth. The "market" signaled demand for more scale, and the market delivered.

Unfortunately, as with the market delivering hash power for Bitcoin mining, the market delivered an unqualified product. The metrics of scale went through the roof, but the purpose of the chain was sacrificed.

And it's not these chains' fault.

Customer demand and a desirable chain are good things. When something is scarce, like blockspace, demand is a sign you're doing something right. If more people want to come to your chain, it will get crowded, but it's only because you built something worth coming for.

The hard part is staying focused on the goal when demands for throughput begin to rise.

Let's go back to adding more lanes, but instead of asking what the traffic situation is and how to relieve it, why not ask what the living situation is like and what the goal is.

If a community builds a nice, walkable city with good

schools, eventually lots of people move there. It fills up, house prices go through the roof, and roads experience traffic. Demands for scaling abound; your town is seen as NIMBY if you don't want to build that high-rise and four extra lanes.

You may need to think about who your town is for.

The same is true with chains.

Crypto to bank the unbanked. Crypto to provide credibly neutral computation to allow marginalized communities to organize and fight nation-state power. Crypto as a space free from judgement and censorship where you can exit with your money. It sounds great. But somehow every chain ends up full of grifters and finance. The new actors realize that 'judgement-free' means there are no financial regulations. They create Ponzi schemes and experiment with new financial tools that would be frowned upon in traditional finance. The system gets crowded, and no one who was initially using the system can even use it. Calls for scaling increase, and the walkable neighborhood is sacrificed so the original tenants can afford to live there.

The problem isn't that these systems don't scale; it's that any system has limits. There is no database that you can infinitely store things on for free. Priority needs to be given in some manner. By prioritizing markets as the method to distribute resources, this simply gives it to the person willing to pay the most at any given moment. It's like distributing access to the roads by having an auction every minute. One day it might be affordable, but if one big business moves in, then you might find yourself confined to your house.

credible neutrality as open blockspace

All of this said, we arrive again back at the myth of credible neutrality. The failure to articulate a specific set of use cases or users in the name of fairness and openness leads again to capture and indifference. All distribution decisions are given to markets under the false belief that allocation via anonymous auction is the only fair way to distribute resources.

"Everything should be open to all."

But as the system's indifference towards you becomes apparent, you become indifferent toward the system. If resources are in any way scarce, users are competing with one another. Whether it's simple market dynamics that price out lower-margin use cases or, as is the case with stablecoins, some use cases can even threaten neutrality itself.

We've seen the simple scenario play out over and over again:

- The founder makes a system.
- Early adopters hold true to some principle and build around it.
- New users come in and don't hold the same values.
- Eventually there are way more new users than early users.
- Early users are priced out or no longer want to be there.
- System no longer holds true to founding principles.

This form of capture appears to be almost inevitable. We see it in crypto, corporate buyouts by private equity, and even democratic systems yielding to corporate lobby; someone sees the potential to utilize the system more than everyone else, influence accumulates, and capture occurs.

Every system wants new entrants. It wants to be able to change the leaders so it can stay relevant and innovative. It wants to stay open to all, but no one wants to lose the reason the system was created.

It's an impossible task.

We've backed our way into a modern version of the Ship of

Theseus paradox. We're replacing every plank and beam of our ship one at a time, all the way until none of the original remains, and now we find ourselves wondering if this is still the same ship. At what point in the transformation did it cease to be the original?

If the original developers wanted to 'bank the unbanked,' but they gradually sell to hedge funds, and early users who believed in censorship resistance are replaced by grifters engaging in regulatory arbitrage, when is it just something different?

And unlike the ship, the code doesn't even need to change. For Bitcoin's early developers, the story of a peer-to-peer cash was replaced by investors who wanted 'digital gold.' It was the narrative, not the code, that changed. When Ethereum's vision as a tool for revolution and subversion was overtaken by defi traders and gamblers, the actual usage is now entirely separated from the vision. The ship is still sailing, but no one can agree where it's going.

enforcing something

To give an example of a valid goal for a chain, it could try to be a currency for those disenfranchised, to bank those not served by the current market systems, to reduce costs by removing legally enforced intermediaries, or even to help those not taken care of by their local government system. These are clear goals.

But this is just part of it.

Scaling without principles is a race to the bottom. If your goal was banking the unbanked, you actually can't let big finance on. It becomes a race to stablecoins and high-value transactions. And this is what open markets do. They auction off your system for the highest amount. If you don't have a purpose for what the blockspace is to be used for, the purpose is the highest bid. And once those users come on, they will scale the chain how they see fit. Maybe they want more throughput, or maybe they make money if there isn't as much throughput. But you gave up that decision. The chain can succeed by the definition of censorship resistance and credible neutrality but by the original goal, be an abject failure.

None of this is new or unpredictable. We faced these same tradeoffs when scaling Web2 applications. The early internet cypherpunks closely resembled the early cryptocurrency adopters. Many were even the same people. Philosophically driven, they leaned on the internet's ability to disseminate information and connect us globally, all while allowing us to be anonymous and free from the censorship of traditional media outlets.

In both cases, things didn't work out as planned.

As we started building, we quickly got visions of scale. Websites that hosted not just personal videos, but ALL videos. Once we didn't just want 100 visitors but 100 million users who demanded faster performance, self-hosting websites at your office just didn't cut it.

Crypto is following the same path.

Over time we'll figure out what pieces need to scale and which ones we don't. We'll slowly realize what needs to be globally verified and what doesn't. Maybe it's currencies. Maybe it's some form of social media or communication that we want to have. Maybe it's all of finance or property records. It's hard to say which pieces make sense to be globally verified and immutable. But it can't be everything. The tradeoffs are already too apparent.

Leaving the decision of city planning up to the highest bidder does not lead to a desirable neighborhood. It might mean no more row homes. It might mean chain restaurants and highways. It might mean data centers and defense contractors instead of farms. If you have values, you have to enforce them. If you have a vision, you have to design for it.

Yes, blockchains are databases, but they are databases we use to coordinate with. They were made for a reason.

These decisions are not technical problems but inherently political choices about values. We continue to try and optimize the system to scale and serve everyone, but it continues to miss the mark.

We're hopelessly trying to add lanes and wondering why we can't find a city we want to live in.

~

"Welcome gentlemen," said Bennet as he opened the meeting. "Thanks for being on time-ish. I think we have all six of us here, which is awesome. Anything before I click record?"

"No?" asked Bennet as he looked at the boxes for signs of activity. "Cool. I'll start with formalities, and then we can just talk about the proposal, but I think we've discussed quite a bit offline, so shouldn't be too many surprises."

"This meeting is now being recorded," said a robotic woman's voice.

"Hi everyone, Bennet here. This is the Dojima DAO call and today we'll be doing the votes on the tokenomics upgrade as well as on two grant proposals. Are there any questions or comments before we begin?"

Bennet was the only delegate on the call with his camera on. The other muted black boxes displayed only their Discord handles.

"We could probably use a summary of the upgrade for the call if you can," said the box labeled saladToes.

"Sure, Josh, want to do that?" Bennet replied.

Josh turned on his camera and then proceeded to take ten seconds in order to share his screen of the governance forum post explaining the upgrade.

He read verbatim from his post's TLDR on the site: "This upgrade will initiate a one-time token inflation of 400 percent to be distributed according to table 1 in the proposal. Seventy-five percent of these new tokens will go towards liquidity incentives on the chain over the next year. Fifteen percent of tokens will also be used to create a sustainable foundation for a grants program."

He paused as he switched his screen to the two grant proposals.

"Umm, we have one grant for $25,000 to get an oracle provider for our chain, and then just another $100,000 grant to NodeTopia for operation of an RPC and development of an open-source explorer for the year."

"Thanks, Josh," said Bennet. "Are there any thoughts on the proposals, or should we jump into voting?"

One of the boxes unmuted. Bennet tried to do his best to maintain eye contact with himself on the screen since they would be putting the video out to the public.

"Hey Ben, Shep here. So, you know, I just want to chime in that I'm not for the upgrade and the grants. I mean, you and I've talked and I post. You all know how I feel. I know I've gotten a grant for the pool, but I think we can build our own tools. It's just a waste. I'd rather see just less inflation, to be honest. But I think I'm outgunned here with the votes, but I needed to voice the thought on the call."

"Thanks, Shep. Listen, we get it, and this was a move a long time in the making. We've tried to just let you guys and the community do it, but the tools aren't being built to what a lot of new users are hoping to see. The entire space is moving to where they expect these things to be polished and professional products, and we have to make sure we deliver on it."

Bennet paused.

He felt too much like the leader of the system. If he told them to vote it down, they all would. They didn't have an opinion; they just trusted Bennet to get the vision right. Wasn't this decentralized governance thing supposed to run itself? It felt a lot like no one wanting to take control. The whole facade was more like some sort of odd regulatory workaround to justify him calling the shots.

He took a deep breath. He knew why it was like this; he just couldn't talk about it. The token had been down like the rest of the market since launch, so this was a nice little last-ditch effort to get things going. The team had runway and tokens for another year, but they needed something to take, or else none of this would matter.

"Let's just do the three votes at the same time, and we can break down if needed," said Bennet.

"All in favor of all, say 'Aye.'"

Shep was silent, but "aye" came from the other four boxes.

"And an 'aye' from me," said Bennet. "Quorum has been reached."

He felt like he should hit a gavel or something. Instead, he just gave a half smile.

"Exciting times, everyone. Really looking forward to seeing this through. Talk to you next month when it's live."

6

THERE IS A REAL WORLD

If you are neutral in situations of injustice, you have chosen the side of the oppressor. If an elephant has its foot on the tail of a mouse and you say that you are neutral, the mouse will not appreciate your neutrality.

-Desmond Tutu

Jim and his two best friends were finally about to head in. The three 15-year-olds had spent the last 12 hours on the side of the strip mall. They had waited all night for the theater to open.

The Empire Strikes Back was finally here.

Their supplies of Doritos, bottles for pissing, and a two-liter of Coke had lasted. The morning sun was getting hotter and they had finally made it to the front of the line. Even though the theater was using all available screens for the preview, they knew that many in the line of several thousand people would be forced to wait until the second showing.

But not them.

When they got to the ticket booth, they paid the usual fare, $3 for entry. The boys didn't get seats together as they were some of the last ones to file in for the first screen, but it didn't matter. They would be among the first in the nation, especially in their town, to see the new movie, and they couldn't be more excited.

Pricing mechanisms are choices. And they don't have to be perfect.

Jim and his friends would never be able to compete at an open auction for the rights to see the movie first. Luckily for them, the market was not yet efficient in terms of financial gains for the theater. The mechanism by which the distribution happened was not that of money but time.

Financial markets are just a different distribution mechanism. They are one of many choices for allocating scarce resources, and each option has distinct winners and losers. The movie tickets could have been distributed by an auction or lottery. By giving them to the person with the most money or randomly to a given entrant, the system could have optimized for maximum profit for the theater or fairness without the waste of needing to stand in line. Either one would have had its benefits.

Distribution by queue was the method chosen, a path which led to long lines but a sense of meritocracy and general fairness by those who won (or didn't). None of the mechanisms are necessarily more "natural" or "neutral" than the others. Society picks different methods for different purposes.

An auction for kidney transplants, for instance, might efficiently match organs to those with the highest willingness to pay, but society has collectively found out that this distribution method produces morally intolerable outcomes, so we use waitlists and medical criteria instead.

Education in the US is distributed by geography and age. If

you're of a certain age and live in a given area, you get a seat in the classroom. Of course, there are inefficiencies in this method, but it was a deliberate choice to make something a right rather than a commodity.

The same thing can be allocated by market price, government allocation, lottery, queue, professional judgement, democratic vote, or social status. Each method optimizes for different distributions and views of equity. Monetized markets optimize efficiency in financial terms, queues for patience and free time, government rationing for political criteria and favoritism, and professional judgement for expert-defined merit.

When distributional methods are chosen, they will always preference some individuals over others. Queues like our movie example tend to favor those who are able-bodied and able to wait in long lines. Financial markets, on the other hand, are biased from the start to systematically favor those with certain accumulated capital. Each design has its own assumptions about which forms of social capital should be rewarded when resources are scarce. Each design has differing views of what "fair" means.

Most of the time with crypto, protocols trend toward using financial markets, as they avoid the difficult problems of politics or human judgement in the decision-making process. Purely financial markets are fair in the sense that no one is directly in control, but the outcomes are usually anything but equal or even ideal. As with the ticket example, auctioning off the tickets would have given the theater more money, but would it have been ideal? Unfortunately, in the digital world, queues are not an option when one person can pretend to be hundreds and then just auction off spots outside of the theater. Strong identity and reputational measurements can help, but they unfortunately run counter to the goals of many cryptocurrencies. Without some sort of centralized actor enforcing the queue, the open auction becomes the preferred option by default.

These meta-structures, or values held by the participants, often determine the method. The principles of neutrality or censorship resistance are upheld at all costs even when revealed that they are incompatible with the outcome of the distribution. This only becomes tenable to uphold as the method itself is entrenched as the goal in some form of religious right to meritocracy or immutable contract law over an actual society.

We see this in traditional business, where the most common form of financialization is in the structure of corporations. It's no surprise that these setups often lead to organizations forcing financial efficiency on every aspect of the business in order to generate the most return for the shareholders or executives. It bleeds down throughout the organization. Current trends in movie tickets have shown that we've pushed even further down the line toward profit maximization. From giving the seats different prices based on show times, location in the theater, or amenities, these are all forms of adjusting markets to make them more efficient; they seek to charge the exact amount someone is willing to pay.

This helps the owner from a profit perspective, but many times, customers and citizens in a given society don't want the distributions resulting from hyper-efficient, profit-maximizing markets. Would zero line but $50 movie tickets be a good thing? Maybe, but it depends on who you ask.

Another example is housing.

In Singapore, housing is distributed by both markets and government. The majority of Singaporeans live in homes built and sold by the government. Started in the 1960s, the government saw a need for affordable housing, so they delivered it. The system gives subsidized loans to first-time home buyers, citizens, and families. Resales and rentals of the units do happen, but it's highly regulated. The obvious drawback is that the system pushes certain views around who should own homes (i.e., families and citizens), but it's a tradeoff made to benefit the citizens of that country. Whereas most markets struggle to provide affordable urban housing, most view Singapore's design as one of the most equitable housing distributions in the world.

In America, housing is distributed based on auctions, but with an added component of loans. The method chosen is an auction that's split between those with capital (previous homeowners or wealthy individuals) and those willing to spend future income (long-term loans). Additionally, long-held views on ownership of land and the right to live somewhere forever leave distributions rather sticky for decades, or even centuries, for some properties. Since building is also handled by a frustrating combination (the private market funds and builds it only after the government approves it), there seem to be problems of never having enough housing. This has led to certain distributions in housing that some, particularly younger generations, find inequitable.

Going back to the drawing board on the issue of housing, it's safe to say that if we found a planet that had the current housing supply and we needed to place the current population, we might do it differently. Moral views of equity and meritocracy aside, a more deliberate distribution might yield something along the following lines: "Young (starting) and larger families should have bigger houses, older and single people should have smaller ones, and those that do the most for the community (teachers, doctors, and community leaders) should get the first pick..."

Clearly that's not where we ended up.

The problem is that "who deserves" or "what's fair" are subjective questions that don't scale well when creating a universal policy. Additionally, picking winners and losers is a hard conversation that's subject to capture by decision-makers, so systems just auction it off and accept the outcome.

But it hasn't always been this way.

Throughout most of human history, unspecified social norms governed human interaction through direct, face-to-face enforcement. The broad consensus of those around worked well enough. Either we do what we've always done, or we all agree on some action. In the most extreme of examples of mob rule via consensus-based judgement, one needs look no further than the Salem witch trials. If a community agrees she's a witch, it's pretty clear she's a witch to all stakeholders, and they hang her.

It's definitely not what most would call a just system, but it is adaptable. Change toward more rules-based systems came as humanity scaled itself beyond just in-person interactions.

Whereas our local community can come to decisions since we all know and trust one another, if a distant town accuses one of our own of being a witch, the feelings might not be as mutual. As a result, we created laws and processes to limit these mob mentalities. As control moved beyond those we trust through our own experience, the desire to live in a system of rules began to grow.

Over time, however, the rules grew and grew until most areas of our lives were captured by legal systems or markets and not based on personal relationships or understanding. Now we realize we don't like the resulting distributions, but feel trapped by our own legal system. We realized that a system of codified rules had its limits.

They remove the nuance in life. Changing circumstances became difficult to handle and the markets themselves were gamed from the outset.

Knowing the system didn't enforce the socially accepted view but rather the exact wording enforced by the law, loopholes were found. Wealthy individuals found ways not to pay taxes. Law enforcement was captured. Rules were applied unfairly to different subsets of a population. Corporations "obeyed" the law while completely ignoring the spirit of its creation. As such, we find ourselves in our current situation, looking for ways to add context, humanity, and common sense to the rigid law.

is efficiency the problem

Participants in market systems have become increasingly good at gaming the target metric. If the goal is profit, the "market" will figure out the best way to make the most money out of it. We saw it with Bitcoin. Bitcoin gives rewards for hash power. As a result, people scoured the earth and even invented new ways to deliver the input.

You can see it with other goods too. Take a well-known example of airline tickets. If it's realized that the market pays for only the cheapest seat, the airlines will optimize for the cheapest seat. The included checked bags, free snacks, comfortable amount of legroom, and even the safety of the plane are often sacrificed to deliver the metric.

And this is a feature of markets. They take advantage of what the consumer is paying attention to. Competing entities iterate through unchecked assumptions to remove all that is unnecessary. If the customer isn't paying attention to seat quality or snacks when making the purchase, it will be sacrificed in the name of profit.

Markets assume perfect information, but the problem is that information takes time to disseminate. Eventually consumers will push back and attempt to get the desired input. Eventually we'll rally our congressmen and push through a vote, but it'll be years down the road. Eventually we'll ask for better quality, but the airline seats never quite feel like they did in the past.

This process is also exhausting. The market will continually underprovide and search for ways around. As soon as you pay attention to one thing, another is sacrificed. It's a game of economic whack-a-mole that constantly drives the system to its own demise. The incentive structures force the consumer to care about every little thing, and it's literally impossible.

Efficiency is the problem. It's an even greater problem if it moves quickly. If a market gives money for X, there is a bounty to deliver X in the cheapest way. For a while, as with Bitcoin or airline tickets, the system works. The market delivers a good experience for X based on what social expectations would assume the market would deliver. The market creator doesn't need to extract because he's the only game in town.

But it doesn't last. Eventually the race to efficiency kicks in.

The lines for movie tickets are gone in exchange for seat-by-seat pricing. The extra legroom is something you need to pay for. The hash rate is supplied by stolen energy and cartels. The "efficiency" destroyed the utility.

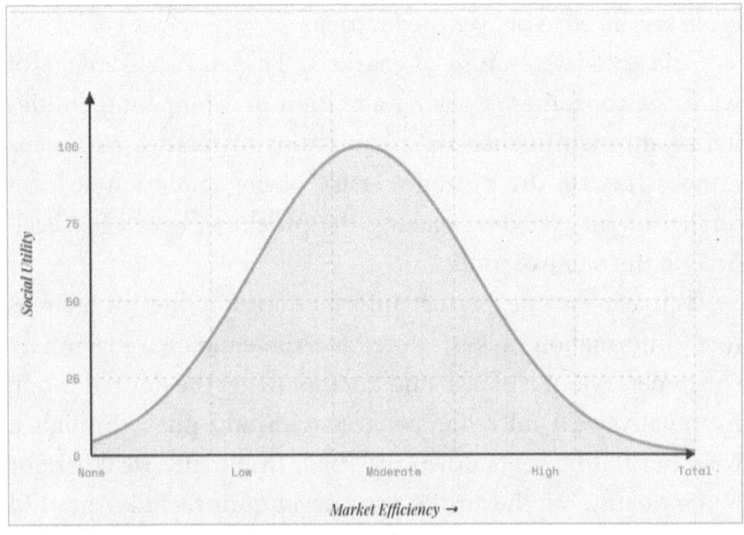

Social Utility as a Function of Market Efficiency. Author's own figure.

Our religious adherence to markets allows for capture. Over time we went from the market being a chosen distributional method to it being the only method. We're not deciding distributions as a society; we're making the market efficient under the belief that this efficiency will solve all of our problems.

Trickle down utility.

The game to target the metric of profit as the only correct path.

And it's this fixed structure that's the problem. It's not adaptable to what society wants or if society changes its mind. Pure market proponents have created a tautology where the outcome of the market is what society wants because the process can't be wrong and the process can't change.

And this is why crypto is dangerous.

One of the primary uses of crypto and blockchains is the codification of markets. Success is finding a market structure that produces a desirable outcome and then freezing it.

The distributional method, locked in code.

In reality, the distribution is sacrificed to the efficiency game. When creators state outright that the gaming of the system is the goal, and they'll never step in, people listen. They game the system and they do it quickly. Whereas in most markets, it takes time to increase efficiency, in crypto it's accomplished at hyper speed.

Crypto's digital nature and financial use cases are the technical reasons it can move quickly, but the underlying factor is the immutability. Whereas in traditional businesses and markets, it takes time for society to become numb; crypto is numb as a feature. If you went back to the 1970s and introduced airline tickets with baggage fees, uncomfortable seats, and fees at every turn, you'd likely be labeled a scam. If you told everyone in line at The Empire Strikes Back that for the first day, tickets would be $50, they'd break down the door. The market can't force itself upon a society unless the customers accept it.

But now the process itself has become the goal for society. One person's high fees are another person's profit. We're back to being members of the syndicate.

The veneration of the market over the actual outcome. The acceptance of the market equilibrium over social consciousness.

In crypto, there's a term for the community of a system: the social layer. It's the larger circle of stakeholders that give a system its value. The social layer is the humanity in the system. As with the discussion on forks, the social layer is the group that collectively can act outside of the rules. In traditional government and law, this would be the social change of a system's rules. If you have a revolution or even a democratic vote to update some outdated law, this is the social layer moving to voice itself.

Although the idea exists in crypto, there's a current movement within the ecosystem to make the social layer unnecessary. The ultimate goal for many is that the systems operate purely as machine code. They are launched and never change, ready for anyone to optimize or interact with.

Unfortunately, many are realizing that these systems don't work without a society. The law needs the social layer for legitimacy in the same way the social layer needs the law to prevent mob rule. Things change, and code needs to adapt. Unforeseen situations arise, and protections need to be put in place. Markets need a social consciousness. Without one, there is nothing to prevent the markets from destroying themselves through efficiency.

penny auctions

My favorite example in American history of the social layer is that of penny auctions. During the Great Depression in the 1930s, America experienced a stock market crash, banking crisis, and global drought all at the same time. Economies around the world were devastated. Unemployment skyrocketed. As a result, many farms and homes were foreclosed upon due to an inability to pay their mortgages or property taxes. As is typical with home loans, once the borrower defaulted, the bank would take ownership of the property, and then sell it to recoup their loan. In practice, this meant auctioning off the house to the highest bidder.

But these weren't just numbers. These were real people going through real pain due to no fault of their own. As a result, the social layer stepped in.

Seeing this removal of homeowners from their land as an injustice, the people of these local communities fought back against the banks. On the date of the auction, neighbors and friends would arrive with pitchforks and axes. Once the auction started, the homeowner would place a bid of a penny, and then using intimidation, the axe-wielding townsfolk would physically threaten anyone else looking to bid on the property. As a result, the homeowner got the property for a penny and their debt cleared.

Although the homeowner did default on their property and, in the eyes of the law, they were due to be removed from the land, society at large saw this as morally wrong and changed it. Some, especially those looking to protect the bank, viewed these actions as lawlessness or anti-market, but the practice of penny auctions is generally looked at in a positive light historically. When push came to shove, looking out for your fellow man took precedence over adherence to a law.

The codified rule (one must pay off their mortgages, or the

bank gets the house) was created for a purpose: to promote loans. The idea was that if banks and investors had a greater expectation of repayment and could recoup their investment if it didn't pan out, they'd make more loans. As a general principle, it makes sense; we want people to be able to get loans using their property as collateral so they can buy homes sooner, use the money to build a business, have more kids, or whatever else they may want to do. It's a good law, and most societies recognize it as such, but in a depression when people were starving, the purpose of keeping loans flowing came second to making sure people had a place to live. When the law was written and the mortgage taken out, it wasn't foreseen that the inability to pay would be due to no fault of their own.

Looking back, would an automated auction of the house be a good thing?

Would a more streamlined default process help society?

need for humans

The essential function of any rules-based system is compression. Whether it's literal codified rules like crypto or just the adoption of markets to substitute value-based distribution, rules help to distill the complexity of humans into understandable notions. This enables us to cooperate without requiring personal knowledge, information, or even the known intentions of the other party. If we can condense our entire interaction with another human to rules written on paper or code, then we can coordinate at a larger scale, as both the parties and the social layer now only need to look at the agreement in question to see if it's going as planned.

But there always is nuance.

There are always things you can't say in a contract or things you don't think will happen. It's for these reasons we almost always have a backstop of subjectivity. We have a recognition that human judgement is the final say because not everything fits neatly into the box.

A "jury of peers" is a recognition that you can't foresee all outcomes in a written contract. It gives confidence that "people like you will interpret the nuance." A democracy is based on the idea that humans can place some subjective decision-making in the process of updating the leadership of a government. Even on a smaller scale, there's an option to talk to the manager.

In fact, one might even say that for most systems, the rules are a mere formality. For most governance systems, they aren't even necessarily as concerned with "what's true" or "what does the contract say," as they are with everyone understanding the process and going along with the outcome. The rules need to make sense, but the system in totality needs to signal to people that it's fair, makes sense, and is not just restrictive red tape or oppression by the party that forged the contract. It's a performative function to signal that a revolution doesn't need to take place.

a tale of automated decisions

Enamored with the automation and streamlined efficiency of the technology, a few of the citizens convinced the tech-forward mayor to push the boundaries: they planned to automate the mayoral office and give all decisions to their custom AI.

Purpose-built for this town, the computer program incorporated all of the latest training data and the best models for governing a town of this size. It aimed to be the most efficient government in existence, promising to lay off the vast majority of unnecessary paper pushers. It could plan budgets, send out updates, and even issue taxes to pay for new or unexpected expenses. It was truly a sight to see.

A few months later, a new issue arose, calling for the locality to figure out how to pay for a new road.

There are two proposals: one is to tax the rich guy who owns 95% of the town, and the other is to do a "fair" tax that makes each person pay $1,000. Eager to try out their new system, the citizens do their futuristic duty and plug the data into the AI, and it starts to churn. A few seconds later it says, in its best calm, old man voice, "We're going to distribute taxes evenly so that no man is set apart. The road is all our burden, and punishing our investors and savers would be a poor choice for our long-run security."

Now the question is, what happens next?

As you can imagine, the citizens throw the computer out the window. The whole AI government thing would be looked at as a bought and paid for scam by the rich guy. His connections to the consultants, whether real or not, would be the talk of the town. And of course, the town is likely to have a new mayor after the next election.

Both the representation of fairness (i.e., a system that cannot be captured) and the feeling of fairness are equally as important. If everyone feels that everyone else feels the same way they do, then social action can happen. Only when we're all on the same page do revolutions take place.

As value and markets are subjective, the answer to "what do we do next" or "what are the ideal distributions" is rarely found by letting the system work itself out. Creating a fixed set of rules, even if automated in their intake of new knowledge, can never work for all decisions. The context required and the nuance of human feelings regarding fairness and equity complicate it to such a degree that the process itself must be subject to their input.

avoidance as an impractical strategy

Market efficiency is not a panacea.

In fact, it's the opposite. Just like a board game, once someone takes it too seriously, it's no longer fun to play. One person's drive toward efficiency in winning the game will lead to everyone else just moving to a new game.

Distributional systems require social acceptance, and those systems that attempt to remove social input cannot deliver on the promises they hope to keep. The real benefit of markets is in changing distributions toward a given end. The goal is to enjoy seeing a new movie. It's having housing for people to live in. Society is anything but neutral regarding the outcome.

~

Marco took a long drag of his cigarette. He stood at the high-top table scrolling through his phone.

Bennet casually took his place on the other side of the table.

Marco looked up. "Bennet, right? Thank you for getting me in," he said with a nod.

"Of course," replied Bennet, "any friend of Diana's is a friend of mine. I'm just glad we could meet up."

"The rooftop events are pretty hard to get into for those of us without tokens," he said as he pulled out another cigarette, gesturing toward Bennet.

Bennet waved it off.

The men quieted as a gust of wind hit the venue. Bennet took a sip of his drink and looked out at the lights of the city.

"Not a bad view at all," Marco said.

"Definitely, but yeah, we use these guys for our RPC. They gave us plenty of tickets, so really don't think anything of it," said Bennet. "I hear you're an auditor though, right? How'd you get into that?"

"Long story," he replied after sipping his drink. "I was a trader at a bank for a while; you know how it goes."

Marco had a European accent Bennet couldn't quite put his finger on.

"Now I'm with ChainChain Security as an auditor. I've been doing research on the security of defi protocols, and I like your design, so I wanted to talk."

"Ah, sweet man, what kind of stuff do you look at?"

"I'm primarily looking into how blockchains can stay decentralized. I've formally verified models to show that without completely centralized trust assumptions, they need very long finality windows to make themselves secure," said Marco.

"Right," said Bennet, who paused with a smile. He looked in the other direction at the beach while he thought about the statement. He liked Marco already after the question. "Yeah, I think that's correct. It's super hard for any of this to work."

"But then why is everyone bridging and settling instantly? You guys are the only ones with any settlement delay, and it doesn't seem like any of this will work under financial stress. Am I missing something?"

As Bennet stared at the table thinking, a gorgeous girl in a black dress came and offered them refills. They gladly accepted. There weren't many girls at these events, so the men felt like it was almost a treat. She smiled at them but clearly didn't speak much English.

They caught each other staring as she walked away. They both smirked to acknowledge the situation.

"Yeah," said Bennet, "I think you got it, but I think you're overoptimizing on what we wanted a few years ago. The market is pushing for more speed and interoperability, and we're forced to place the security on top of the issuers. The world is going to be trust at the endpoints, not trustless from start to finish. It's not the cypherpunk dream, but it can actually work for finance in the modern world."

"So, you agree the big chains are just pretending?" said Marco, nodding to a sponsor poster.

Bennet let out a small sigh. That was probably the first time he'd said it out loud. This level of honesty was refreshing.

"Completely," said Bennet. "We're like the only ones out there even attempting decentralization anymore. But we can still get there as a space. I think one or two big hacks and a few regulatory crackdowns in the space will show people that decentralization actually matters, and the pendulum will swing back this way."

Bennet turned to look at the event. Marco mimicked the gesture. The scene of crypto founders making deals and planning their nights on a rooftop was inspiring.

Things were moving, and they felt like a part of it.

7

TOKENS – COMMODITIZED FINANCIALIZATION

"Wealth that is stored up in gold is dead. It rots and stinks. True wealth is made every day by men getting up out of bed and going to work. By schoolchildren doing their lessons, improving their minds."
-- *Neal Stephenson*

Widely considered the first example of a multinational, public company, the Dutch East India Company issued its first shares by charter of the government. Rather than fund ventures across the ocean one ship at a time, these ornate certificates represented a claim on future profits from all voyages to be had by the company.

At first, the concept seemed like an absurd game of trust. How on earth could a piece of paper be worth anything if the ship could be at the bottom of the ocean?

But they took off.

Citizens from all over the city could purchase shares. None of them would ever board a ship or explore the Indies, and yet all of them could now take part in the potentially unlimited gains represented by a venture.

The certificates traded in the courtyards, coffeehouses, and taverns of the city. Speculators and investors alike began to play the game. Everyone who thought they could predict the successful return of a voyage could now buy and sell shares on the open market. The demand for both merchants and investors took off.

The effects were felt in ways no one anticipated.

Within a generation, the Dutch Republic was transformed from a small nation of fishermen and farmers into an empire. Ships bearing the VOC emblem became more numerous than those of the Spanish Empire. The flow of spices, silk, porcelain, and coffee accelerated. More significantly, a new financial tool was born that would revolutionize the globe by allowing for greater participation in speculative business ventures.

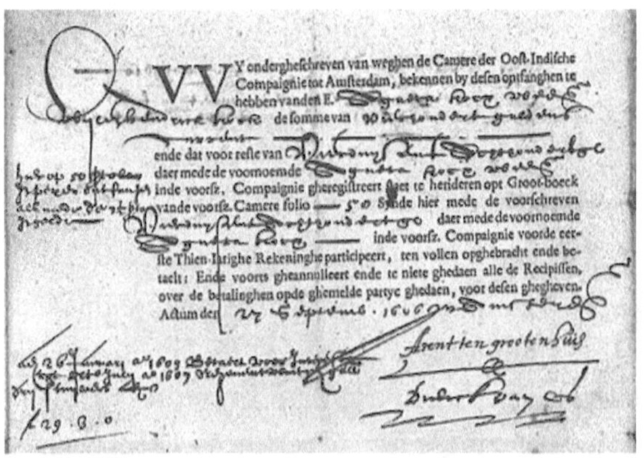

VVY onderghefchreven van weghen de Camere der Ooft-Indifche Compaignie tot Amfterdam, bekennen by defen ontfanghen te hebben vanden E. de fomme van
ende dat voor refte van
daer mede de voornoemde
inde voorfz. Compaignie gheregiftreert ftaet te betideren opt Groot-boeck vande voorfz. Camere folio Synde hier mede de voorfchreven daer mede de voornoemde inde voorfz. Compaignie voorde eerfte Thien-Iarighe Rekeninghe participeert, ten vollen opghebracht ende betaelt: Ende voorts gheannulleert ende te niete ghedaen alle de Recipiffen, over de betalinghen opde ghemelde partye ghedaen, voor defen ghegheven.
Actum den

Share certificate of the Dutch East India Company (VOC)

Or so this was the story.[i]

As amazing as the financial innovation of public shares was, the idea wasn't created out of a belief in free markets and shared profits. In reality, the company was formed as a result of competitive necessity.

Over the previous decades, the Dutch had been outcompeted by Portugal and England in the fast-growing Asian spice trade. In the eyes of the king, competition among Dutch ships was seen as a problem when the real competition should be the foreign nations. In order to prevent this internal competition, the various "pre-companies" (often specific ships) were forced to combine into a single company to which the government granted sole rights to trading with Asia.

The profits for the VOC also didn't appear out of thin air or from honest bartering with foreigners. Throughout Asia, the VOC used its governmental support and military superiority to ensure profits throughout the region.

One of the most notorious examples was in the Banda Islands. Part of modern-day Indonesia, these islands contained the only known source of nutmeg, an extremely valuable spice at the time. In 1621, after trying and failing to secure monopoly rights via a treaty with the locals, the VOC waged war. Dutch Governor-General J.P. Coen led the campaign that reduced the Bandanese population from approximately 15,000 to less than 1,000 through massacre and forced starvation.[ii]

After the genocide, the islands were split into estates to be owned and operated by VOC employees using imported slave labor. The company exterminated nearly the entire indigenous population in order to monopolize the market.

The profits resulting from actions like these led to the VOC becoming one of the largest companies of all time. With its status and navy rivaling the most powerful nations in history, it took the markets of many commodities by force.

The shareholders in Amsterdam did not vote on this or, in most cases, even know about it. The shares helped to fund the actions, but firsthand knowledge was unnecessary for providers of capital.

In addition to the obvious disconnect between the investors and the actions of the executives, the VOC shares themselves also had problems of informational disparities. As they began trading more frequently, professional traders began to take advantage of the commoners. Many of them were wealthy merchants with inside information about ship arrivals and cargo; they manipulated prices through false rumors before the actual information reached the town.[iii] They would short-sell shares to small investors and then buy them back when the actual information emerged.

The financial benefits were realized, but they were far from evenly distributed.

For the next two centuries, the VOC would be a major player in global politics and trade. Over time, dividends began to decline as the costs of maintaining far-away colonies exceeded the profits from trade. The company eventually became too overextended, fighting expensive wars in Asia while paying out dividends to maintain share prices and attract new investors.[iv] Executives, aware of deteriorating fundamentals before the market, sold their shares while publicly promoting the company.

The VOC eventually collapsed in 1799, deeply indebted and unable to compete with English and French rivals. The colonial infrastructure was transferred to the Dutch state.

LIKE THE REVOLUTION that was the stock certificate, Bitcoin, blockchains, and tokenization represent the next evolution of financial technology.

In addition to acting as a currency, people realized early on that blockchain ledgers could be used to keep track of things other than just money. In fact, the idea of digital tokens to represent real-world assets existed about a decade before blockchains.[v] Tokens could represent shares of ownership in real estate, bonds, stocks, shipping crates, inventory, and countless other use cases.[vi]

The goal for blockchains was that these digital database

entries could eliminate the friction of dealing with centralized middlemen. No one would need to keep track of who owned anything (stocks, bonds, houses, etc.). It could all be recorded automatically, and the gatekeeping privilege afforded to the intermediary would be no more.

And so, we issued tokens.

Lots and lots of them.

Tokens that were to be used as currencies, tokens that would be startup equity, tokens that would be ownership rights over software, tokens that represented real estate, tokens that represented crates in shipment, tokens that represented dollars, tokens that represented ideas, tokens that represented nothing, and on and on.

Whether it exists or not, we can make a token out of it and hand it to a global financial system.

observation of value

The general idea was that by introducing tokenized crypto markets to different forms of value, economies could grow and be made more efficient. This theory that formally representing value in financial markets can promote economic activity, be it digital or otherwise, is not new.

In his famous book, *The Mystery of Capital*, Hernando de Soto argues that formal property rights are a necessary piece to growing an economy. Published in 2000, the book claimed that citizens with strong, formalized proof-of-ownership could utilize their land for loans and that this access to financial capital would greatly expand the productivity of a nation.[vii] The theory is relatively straightforward. If I can get loans using my land as collateral, I'll be able to start a business or even access financial capital from improvements I make to the land.

Unfortunately, there's a wide range of outcomes when subjecting assets to financial markets. As seen in the story of the VOC, giving greater financial access to certain aspects of society can have unintended consequences. In terms of land rights, dispossession by outside investors, changing social dynamics, and incentives for corruption can make any resulting financial boon feel like a curse.

Crypto is the modern application of de Soto's formalization. We're not altering the law, but rather digitizing and fractionalizing already formal ownership rights. This encoding of different real-world assets takes them from more analog, often nuanced forms of transfer to the hypermarkets of crypto.

This is the story of "tokenizing" everything we can.

The goal of many crypto projects is to tokenize every asset known to man. The hope is that having the ability to trade, borrow against, and invest with these assets will create economic prosperity. If I can access the equity in my home, my future earnings, this one idea I have, my reputation as a friend, or even my thoughts that might make people laugh, then economic prosperity will be realized.

Now the question isn't whether we can, but rather whether we should.

crypto tokenization

Over a decade in, the benefits of tokenization have yet to be realized. Early on, most developers and projects realized that removing trust was only beneficial up to a certain point. If the stated goals of tokenization are to remove intermediaries and allow access to hyperspeed financial markets, we've found that those goals are not shared across society.

Whereas crypto touts trustlessness and censorship resistance as benefits, many have realized that tokenization in no way requires these features. As de Soto knew well before crypto, you can formalize or digitize legal rights without expressing the assets as "tokens." In fact, many assets and users actually function better if trust is established between interacting parties. This bears true in the timeline of crypto as well. As usage expanded beyond the developers and cypherpunks of the nascent system, many new users found that they actually preferred some intermediation.

To give an example, take the issue of custodians.

Bitcoin and Ether themselves are bearer assets. A bearer asset, like cash, is one where if you have it, you own it. There is no intermediary that can help if you lose it or misplace your private key (or password). This is what self-custody is. But conversely, it comes with its own set of problems; specifically, if someone steals it, they own it. If you lose it, it's lost forever. There are no reversions, no alterations, and no helpdesk.

If someone gains access to your tokens, they can instantly sell them for a different, untraceable asset. The security needed to protect your assets has now risen dramatically.

As crypto companies discovered, self-custody is far from practical for most users. For Americans especially, the "trust" piece wasn't really the issue with banks. Most people would rather have a professional hold their money.

Supply chains faced a similar problem when experimenting with tokenization. The idea was that if two customers were trading back and forth with one another, they could just create a token for each shipment crate or item. Eventually the entire supply chain and all suppliers and buyers could trade and track these tokens accurately on a blockchain. If a payment in crypto goes through, a smart contract could then tell the shipper to send the crate to the buyer. The token then transfers to the buyer as soon as the shipment is made. The whole system could be automated, and those PDF files uploaded and emailed could be things of the past.

The obvious problem exists: what if nothing was put in the crate?

As you can tell, the trust just goes from the intermediary

checking things off to the person marking the box as shipped. Of course, there are real-time monitoring tools and other automation pieces that can make it better, but the bigger question for these use cases came about: "Do we really not trust the person running the database?"

Was that the bottleneck?

The truth is that if two parties actually distrust each other and are trying to cheat the transaction, even trading standardized goods is difficult. Anything related to quality or benefits from repeated interactions is lost or requires extensive monitoring if the deal is to happen at all. In practice, almost all trade requires some form of relationship and shared understanding. Since trustlessness isn't needed (or even wanted), the system could be handled just as well with a database, a few APIs, and some transparency. The issue wasn't the intermediary; it was the other party. The intermediary was in many ways comforting because they at least provided some sort of due diligence on the product and, at best, took on some of the risk themselves.

The claim of crypto to "solve remittances" falls into the same category. To explain, remittances are when foreign workers send money back to their home country. These transactions are usually marked by high fees, costs that crypto companies claimed would evaporate with the introduction of global crypto networks.

Unfortunately, these companies quickly realized that the ledger for recording bank balances was not the bottleneck. In fact, the company holding the cash in the small village and actually giving it to the person was both the bottleneck and the real value-add for users. Crypto found that to eliminate those fees, their tokens would either need universal acceptance as money or someone would actually have to give people cash in each small village.

A siloed system for messaging across the globe doesn't help.

Real stuff has quality issues, monitoring risks, and lots of other problems that require context, caveats, and interactions with actual people. These systems, like other technologies, are subject to the social landscape to which they're deployed. If the larger systems and participants are ruthless and cutthroat mercenaries, no amount of technological efficiency can coordinate them into harmony.

For these reasons, the only sector that crypto has found product-market fit in is finance. Not mortgages or loans, but rather a direct, siloed finance that interacts with other digital assets. The kind of unregulated finance that is the ultimate standardized good. Previous financial markets had gates. If you wanted to launch a new derivative product, you'd need the blessing of banks and regulators. Crypto found a new way. Everything is converted to a standard unit, and besides egregious violations of the law, it's devoid of human intervention.

financialization as a parasite

If the goal of tokenization is the introduction of things to financial markets, the next question needs to be what the goal of this action is.

Finance itself generates no value.

It's not a standalone sector that we should look to grow. Its only relevance and utility are achieved as a complement to other industries. A large financial system is only justified if it can help to grow other sectors of the economy. It should spur investments, create loans to build out infrastructure or research, and even redistribute capital to help create a more equitable distribution of housing or college tuition. If the other sectors are shrinking while finance is growing, this is a negative for the system as a whole.

To understand at a high level, take an economy with a GDP of 100B. If the financial sector is 10B and it grows to 20B, but the total GDP only grows to 101B, you actually shrunk the real economy (agriculture, manufacturing, education, healthcare, etc.). A growing financial sector is negative if it's not contributing at least as much growth to the real portion.

It makes even more sense in a small community. If the town has 10 farms, a shop, and a restaurant, what benefit would accrue by adding a banker? If he or she can help grow the business or reduce risk for the owners, they should be seen as a positive. If they extract five percent in interest from each of the groups and the total output stays the same, society would have been better off just making him another farmhand.

Finance helping grow the economy in monetary terms is just the first step. Innovations can increase the financial wealth of society while ruining it from a social or human perspective.

Even if the numbers do work out, how certain introductions of finance affect an economy is just as important. As shares in the Dutch East India Company illustrate, financial innovations can affect society in countless different ways. They can redistribute labor from being idle to working on new companies and scientific discoveries. They can pool investments to enable ventures which were out of reach due to capital constraints. They can also encourage genocide, war, and transfers of wealth to middlemen or extractors.

So where does that leave crypto as a purely financial technology?

We have to look at the results. To be fair, there are plenty of positive externalities of the crypto ecosystem. From groundbreaking cryptography and networking research to new forms of organization that help us better connect with one another and even censorship-resistant tools that help promote freedom around the world. There are many examples, but if looking at the monetary value of these purposes, they come secondary to the financialized zero-sum speculation that characterizes much of the space.

The best example of a pure zero-sum game in crypto is that of memecoins.

memes for gains

Originally pitched as the idea that you could own a "meme" or an idea, the general premise was that if you identified a trend before others, you could make money. The promise was that if more people resonated with the meme, it would take off in value and signal potential virality earlier.

The origins for these tokens started off with the idea of being able to invest in things such as music or social media posts. If you think a new song or post will go viral, these companies hoped to allow you to signal that to the broader community. If your friend invests in their new band, maybe you'd be more likely to check them out. If you like them also, you can signal to others by buying it as well, maybe even helping the original artist in the process.

Once the concept of tokenizing every act of creativity took hold, the list of things people could want to create investable assets of only grew. The broader memecoin category was born.

Literally a picture, a funny GIF, a saying, or even a person themselves, these tokens hold no promises of future value or development and often have no team or initial backing. They're just tokens that people can create and hope that other people will also buy because of the idea. If you like Peanut the Squirrel, I'll make some coins, you'll buy a few, and so will others because they think it's funny.[viii] If you like this picture I took of a cat or the sound it makes, you can make a coin that represents it and attempt to sell it on the market.

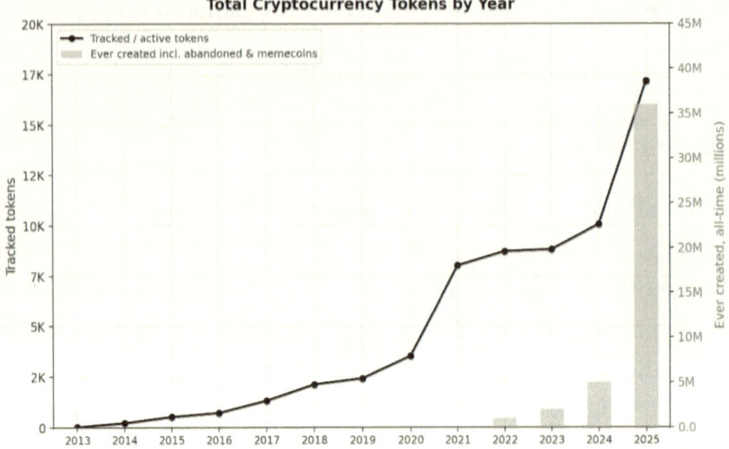

Total Cryptocurrency Tokens Created by Year, 2013–2025. Author's own figure.

Unfortunately, whether well intended or not, these speculative instruments did not unlock some new form of social media or benefit to society.

Memecoins, like much of crypto investing, became a trap for retail. The platforms and a handful of professionals ended up making most of the money. After the first year of the craze, 90% of users lost money. Very few made the life-changing money advertised on social media, and over 97% of the tokens created hold zero value.[ix]

The promise of crypto and financialization creating economic prosperity through memecoins failed to materialize. As has been seen with every tokenization project, the extraction around the token becomes the game itself. The space has seen the same story repeating itself:

- A few projects launch tokens as speculative investments in their promising idea.
- Exchanges are launched to trade these tokens, and so are tokens that own the exchanges.

- Seeing the success of these tokens, platforms are created for launching new tokens along with tokens that own the platforms.
- New ways to bet on these tokens with leverage are introduced.
- Protocols are created to lock up one token to borrow other tokens.
- Tokenized predictions and conversations trade, trying to predict which token will be the most popular.
- Bridges and infrastructure are developed to allow each system to interact with one another.
- Every user, idea, database, or API in the space has a token.
- The original ideas never really work out.

Each new thing adds more "value". The system is worth hundreds of millions of dollars.

But nothing is actually created.

In fact, you might even say that the time wasted and resources drained by employees and speculators are a net loss for society. The concentration of wealth and power tends to tell a story more of extraction than innovation roadblocks. Like much of traditional finance, it was a zero-sum game with negative redistribution effects.

selling the farm

Financialization rarely works out as expected because it accelerates market outcomes. It drives profits to zero, and it hastens the equilibrium distribution of a competitive system. Although many might view this as a positive, how we interact with financialized systems makes these characteristics anything but.

Normally, profits being driven to zero is considered a positive outcome for a market. The theory is simple: a company innovates and makes something new. They can then charge a premium for this, and as a result, get a profit (revenue minus cost). This profit signals to others to come compete, and pretty soon the market is flooded with lower-priced competition, and they too have to drop their price.

This is the dream, and it's how markets work in textbooks.

But unfortunately, reality is different. People like profits. Entrepreneurs and investors won't even try to start a new company in a space that they won't make a return on their investment. So, the question comes naturally: in a system where code can be copied in a second and a copy of the company made by next month, how do you generate profits? If "perfect competition" arrives after a week, how do we still make money? How is the space investable if the game theory quickly arrives at zero return?

The answer we see is twofold: extraction and pivots.

We see this blatantly with memecoin creators and many of the lower-tier investors in the space. They simply launch, do a large marketing push, take the above-market profits for a brief period of time, and then dump it and repeat. Rather than stay and build something of value or grow a community, the profit is in quitting to start something new.

The instant value creation of financialization actually turns incentives around. Rather than promote good investments, it removes any ability to make one. The digitization and access to markets lead to less long-term value (profits driven to zero faster) and earlier liquidity (ability to get out), thus removing incentives for founders and early investors.

For memecoins, the financialization of trends actually leads to less virality. Instead of just searching for enjoyable things, you search for things that are investable. You don't interact with the viral trend; you want to find the next one.

Crypto sees this playing out in the activity of venture capital also. VCs in traditional industries hold an average of eight to ten years. Facebook, for instance, took eight years from founding to exit. Netflix was considered fast thanks to only waiting five years before doing an IPO. In crypto, the average is down to one to three years.[x] This is a massive shift that's evident in the token prices and longevity of the companies.

As projects and investments are shorter lived, it's hard to say that increased financialization and liquidity are helping to build better companies.

Profits are often made when you either monopolize the market or you extract from less-informed individuals. Rather than make small returns from providing a valuable service, companies in most competitive sectors have found that economies of scale and informational arbitrage allow them to earn risk-free profits.

And this is the game.

In theory, markets should be equalizing, and competition should drive profits to zero. So to avoid competition, these companies hide behind complexity. Since crypto and new technology can be difficult to use, asymmetric distributions of information are plentiful, and those who create the interfaces and technology are usually the ones who can take advantage of the situation. At the end of the day, it turns into repeat investment cycles with retail holding the losses at the end.

It's the same story as the creation of stock certificates in the 1600s. Financialization creates vast wealth for some at the expense of others. Tokenization is following a well-trodden path. Financial goals replace societal ones.

If we can all be wealthier on paper, it's best not to ask how. But we know markets don't just make things better by default. Finance is complementary and redistributive. Its benefits should be judged on whether we like externalities or social changes created as a result of their introduction.

Financialization can often add complexity that leads to

abstractions of responsibility and moral hazard disguised as profit maximization. In the case of tokenization, we're abstracting away the end goal of the product. The responsibility of the asset issuer to society is turned into one of generating a profit over an actual product. Asymmetric information is created to exploit and enrich some. The profits entice the next wave of liquidity. The gambling, theft, inequality, and social division are packaged as unfortunate side effects of freedom. In the case of the VOC, financial tools led to literal genocide. For tokens and the next wave of finance, we're likely going to look back with similar disdain.

~

"Hi, everyone," said the girl labeled 'Marissa.' "Lots of newcomers here, so we can all do intros here in a minute. Bennet, have anything before we get started?"

Bennet looked at his Zoom and saw over twenty people. Lots of muted heads with gaming headphones. The team was growing like crazy after the run-up in price, and he was just happy to have someone else running the calls.

"Yeah, thanks, Marissa. For those that didn't know, today we have six new guys joining us."

Bennet continued to clappy hands and applause emojis. "I know it's going to be quite a transition too. We're moving fast, and all these new guys are going to help us level up the protocol. And it couldn't come at a better time too. We have lots to do, but before we get into what each person's going to be working on, I think it's important to all be on the same page as to why we're all here."

Bennet paused and looked into the camera.

He felt like the leader now. He had never managed anyone in his previous job but now had all of these people relying on him and the system he built for a salary. Part of him felt like he needed to give them a purpose and a reason for being here. The other part felt like they couldn't care less, and they just wanted a paycheck.

He didn't know, to be honest, but he had read enough times that mission statements mattered. At the very least, he always appreciated some purpose in his work.

"We're here to revolutionize finance," he said, glancing at his notes.

"Centralized currencies and the big banks are preventing progress, and we're creating a system where anyone can create and use financial tools, and we think that the world can be a better place because of it. People can avoid inflation, they can get around predatory fees, and they coordinate to start new companies and innovate, and Dojima is looking to be the backbone of all of it."

The team stared at him through their screens.

"It's our job to deliver on this. We really have to stay focused, and we can't get captured by the easy way forward. Centralizing our system and becoming TradFi in disguise is not an option for us," he continued. "We have to be better than them too. Not just from a technical standpoint, which we will definitely do, but from a moral one. We're not just building a fintech startup or working at a bank, and I don't want to. We have to remove fees, we have to build in the open, and we have to democratize every aspect of our protocol."

He felt a little like a preacher, but he didn't know how to stop. He didn't like these kinds of talks, but motivating people meant motivational speeches, right?

"I'm super excited to be working with you all on this. I want you to challenge me on this too. Keep me focused and call me out if I'm not being consistent. If you ever want to talk, I'm always available, and even more so, I really want to hear from you all too. I'll be setting up meetings to talk with all of you individually over the next week, but think hard about the why and let me know what can make this a better system."

He finished, slightly self-conscious of his closing line. The call was silent for a few seconds. He got a nod from a few of the guys and a smile from Marissa, who took back over.

"Thanks, Bennet. Now, let's just do our normal dev call, and we can introduce ourselves as we go. I'm Marissa; do the events and manage the docs… yay," she said with a smile. "But I'm doing a rewrite of the liquidity provider sections today, and I'm going to be working on our booth design for Japan next month. So. Yeah. Nice to meet you all. Josh?"

8

PORTABLE VALUE – GAMING, NFTS, AND OPEN REPUTATION

"The technology is just gonna get better and better and better and better. And it's gonna get easier and easier, and more and more convenient, and more and more pleasurable to be alone with images on a screen, given to us by people who do not love us but want our money. Which is all right. In low doses, right? But if that's the basic main staple of your diet, you're gonna die. In a meaningful way, you're gonna die."
-David Foster Wallace

Axie Infinity is the quintessential example of a "blockchain-based game."

Slightly early to a much larger movement, Axie launched in 2018 and sought to revolutionize gaming through "play-to-earn" mechanisms. Their thesis, as well as that of countless other startups, was to give gamers crypto incentives in order to boost traction and create a better and fairer gaming experience.

Anyone with a teenager knows that players dream of the

opportunity to play video games and earn money. You didn't even have to explain the "customer" to potential investors, as it seemed almost self-evident. The further claims of digital ownership and connection to the world of decentralized finance made it a narrative that fit perfectly with the moment of crypto. People from around the world could be playing a game, and there would be real, monetary incentives for being the best. This was the future of gaming.

Screenshot from Axie Infinity, developed by Sky Mavis (2018). © Sky Mavis.

THE GAME itself was far from extraordinary. [i]

Players would fight in battles or complete daily quests to earn tokens.[ii] They could then use these tokens to breed new characters ("Axies"). Since you would need to "burn" tokens to make these new characters, and everyone needed to buy new characters to play, it created a theoretical supply/demand loop. As new players entered, they would bid up prices for new characters in order to start completing their own quests, and the cycle would continue.

In case you didn't realize, the system was flawed from the start. Its economic viability depended entirely on perpetual growth. It was the classic, unsustainable pyramid scheme. Without constant newcomers to buy the new characters, the system would be minting tokens with no buyers.

The crypto community, always ready for a new narrative, rationalized that network effects and real engagement would eventually come to make it sustainable. VCs invested $161 million in Axie alone, and other companies collectively raised billions on similar narratives.

While it was growing, the system worked as intended. At its peak in 2021, over 2.7 million players were playing the game. The trading volume for Axies hit over $1.3 billion, and play-to-earn looked like it had found product-market fit in the best possible way.

But, as you can guess, it was short-lived.

By 2023, most of the players were gone. The decrease in players joining led to a collapse in value for new characters and also the token. As of today, the token is down over 99% from the peak, and the system is down over 95% in terms of players.

but why did it take off?

The end result of a failed Ponzi is nothing new.

In crypto, especially, for every pyramid scheme that blows up, there are 500 that fail to ever make it past a few thousand dollars. The more interesting piece is why it took off and what the characteristics are that we can generalize about all financialization (even ones with sustainable token economics).

The biggest lesson from play-to-earn games is that you get what you pay for.

Companies and investors like to pretend that paying for users will lead to more users of the same caliber they already have. They imagine that their system will just end up marginally better than that of competitors because of the rewards (like airline miles or a free ice cream after your tenth visit).

Unfortunately, what happens in these hypermarket systems is that anytime you pay for an outcome, the market delivers you that outcome in enormous quantities. It will very quickly become efficient and innovate to deliver you as much outcome as it can in the cheapest possible way.

For Axie, its minting mechanics paid for gameplay, and it got gameplay in the form of fake accounts and bot farms. Players would often create tens or hundreds of accounts to circumvent earnings limits built into the system.

But this was just the tip of the iceberg. Professional operations were built to play the game for money, specifically sweatshops in low-wage countries (think Philippines, Indonesia, and other countries with cheap labor). They recruited workers to play multiple accounts simultaneously, often using automated tools to play for long hours in order to gain rewards.

This was probably the biggest reason for the downfall. By failing to fix the problem of bots and fake accounts, they dissuaded honest players from even partaking in the system. Just as no one wants to be on a social media platform filled with bots, games with these paid players are shells of the real thing.

The Axie system still exists today, but in a much lesser form, and the sector of play-to-earn has lost much of its former sheen to both players and investors. [iii,iv]

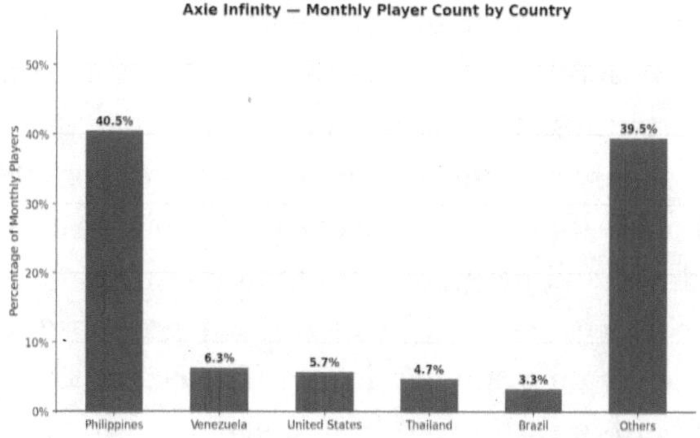

Axie Infinity Monthly Player Distribution by Country. Author's own figure. Data source: Priori Data (2022)

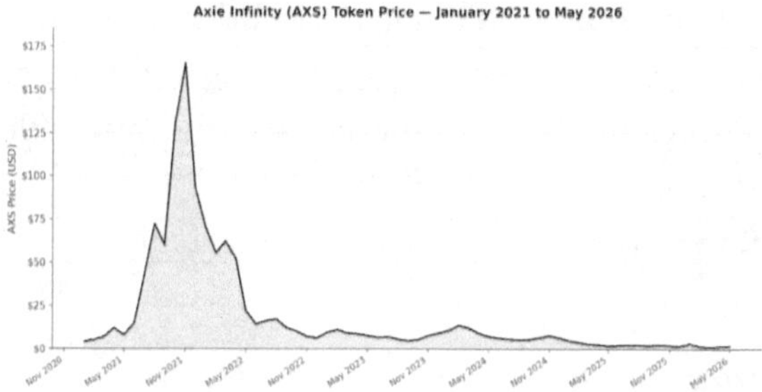

Axie Infinity (AXS) Token Price, January 2021–May 2026. Author's own figure.

non-economic financialization

For some reason, investors and entrepreneurs fail to heed basic economics. Every few years, there's a resurgence of the narrative of financialized gaming and social media. The general pitch is that users create value for these systems that they would like to access elsewhere. Whether by posting on social media, playing and earning points in a multiplayer game, or even creating content, the platform's success is a result of the user's input, so they should get their share. Additionally, if customers have contact lists, followers, friends, or even points of some sort, they're at the mercy of the platform should it choose to censor them or reduce their reach in some way.

Many startups in the crypto ecosystem argue that it would be great if these systems and their data were more decentralized and open. If they were, anyone could spin up a competing front-end and still have access to the underlying data (scores, friends, contacts, etc.), which users had previously spent so much time curating and developing. The classic examples are selling a hard-to-earn sword from a game or bringing a list of followers to a new platform (so building a set of followers doesn't have to start at zero).

The pitch is that these companies are democratizing the distribution of funds and data. By either creating discrete tokens out of every object or just posting the base data on a generic blockchain, the information and the "valuable" pieces exist outside of the game itself.

No more vendor lock-in, no more censorship.

the dismay

The theory hasn't proven sustainable.

As with any business, you can attract users if you pay them. Even specific to crypto, PoW mining, liquidity farming (a popular practice of paying people to provide liquidity on a token), or even hinting at airdrops (free tokens for usage), all of these have proven to generate massive user and volume metrics for a system. As fast as they come, however, each of these systems has proven that these users go away as soon as the subsidies stop.

Blockchain gaming and social networks are no different.

The issue stems from the difficulty of using economic levers to drive non-economic activity. The reason people play a game isn't to make money, and the reason people are on social media isn't to earn a living. The time spent on these activities is largely leisure; it's to enjoy oneself, talk with friends, or stay up to date with a community. Of course, there are some that will chase profits and come to a system for monetary rewards, but for the vast majority of users, they pick the game based on how good the gameplay is. They pick the social media platform based on how good the community is and what it offers.

Financializing the system doesn't make it more desirable.

If anything, the incentives can degrade interactions by altering user behavior. Those "new users" who are corporate sweatshops and AI bots dissuade actual users from interacting with your system, and studies have shown this. As anyone who's seen bot replies on social media can tell you, these users pollute both the composition and quality of the user base.[v]

transiency

If incentivizing people isn't always a good thing, what about the story of leaving with your value? Don't users want systems that allow them to take their hard-earned reputation or social graph with them?

The answer is shockingly no.

Unsurprisingly, individuals do want to be able to leave with their own value; the only problem is they don't want to partake in a system where others can.

Exiting with value has long been a known problem of market-based systems. People and capital are fluid in market systems. If a market opportunity exists somewhere else, the capital and labor will flow there. If a company can make something more efficient elsewhere, they do so. If they can lower the cost by moving production, all else equal, they will.

By making reputations, social graphs, or in-game currencies transferable, the once-closed systems of games or social networks are now subject to market forces. It alters the entire landscape of these systems. For a game, this usually means that if people can leave with their value, they leave earlier and don't stick around. If participants come to make money and not just to enjoy themselves, the entire community and the feelings of those playing the game change. The once collaborative game is now a cutthroat competition that determines certain players' livelihoods. The social media platform for sharing updates and ideas is now gamified to encourage users to farm views or likes (whichever leads to the higher reward).

Money and value don't stay in one system unless they're forced to. When everything is measured and commodified, there is a natural force for capital to go to the area where it is desired most. In the same way water flows and fills in every crack, if there is a return on capital, the market will find it.

If you earn money in one town but want to travel or buy goods from another town, this is transiency of value, and we generally look at this transiency (trade) as a positive outcome for both parties. But the system needs two towns that both benefit from the trade. If one town has something valuable (the game), there needs to be another town that allows you to bring your value in. For a game, this would be another game allowing players to bring the assets over. For a social media network, this would be a new system that bootstraps off of another system's network.

But if you're the new game with distribution, why wouldn't you just mint your own currency from nothing and capture all of the value? If a new game incentivizes a famous player from one platform to come onto their system, this is likely beneficial, but if they allow that popular player to sell his status to the highest bidder...what exactly did they achieve? They didn't even manage to get the famous player on...just a financialized representation of their account. If you try to incentivize new users by saying, "You'll be able to leave with your value," you follow in Axie's footsteps and will largely just attract the kind of users that will play solely to leave with their value.

Conversely, if you're the old social network and already have success, why would you let users leave with their reputation? Don't you want to own the customer?

And this is exactly what we see. The party that has the distribution, whether it's the new game or the old social network, doesn't want to interoperate. They close it off because it's not a beneficial agreement for them.

Sometimes we want things to be fluid and move across borders. Other times, we want you to have to earn it. Some trust and loyalty can be bought, but the majority of it needs to be earned.

values and exit

Friction within markets is almost always viewed as a negative.

In labor markets, the human preference for staying in one job or location prevents smooth transitions in labor markets and can be an explanation for why markets aren't efficient (why wages in the US might be higher than in other countries for the same work, for instance). Removing frictions might help from an efficiency standpoint, but the mobility of people can lead to unintended social consequences. Broken families, immigration tensions, housing prices, and countless other issues are all tightly interconnected to the efficiency of the labor market.

In fact, one could even make the argument that additional frictions might even be helpful from a social cohesion point of view. In the US, citizens are often given preferential mortgage terms (e.g., FHA and VA). Tuition discounts for college exist for in-state students. Immigration laws exist. These protections are seen as normal and beneficial. Very few people would advocate for instantaneous labor movement across borders, and yet digital assets are consistently pushed towards maximum efficiency with little heed towards existing social structures.

In traditional industry, it's well understood that inputs have social costs. Labor, natural resources, and physical capital have long histories in which society battles with the efficiency-seeking market. From environmental protections, labor rights movements, and discussions on inequality and rent extraction, the equilibrium found at any point in time is anything but the result of a maximally efficient market.

Anytime new market structures were introduced to shift the status quo, social repercussions followed. Introductions of free labor markets led to children looking for jobs in other cities, breaking family lineages in small towns. The introduction of free capital movement allowed corporations to move jobs and income to lower-tax countries. Unrestricted imports of goods have led to countries dodging environmental regulations by moving production to countries without them. Market forces without limitations and transient inputs can force the "best" outcome to simply mean the best as represented by a financial price.

In some sense, digital markets promise us perfect globalization. They offer perfectly fluid markets that find the cheapest inputs across the globe. We just need to ask whether it's a good thing.

Friction might slow down innovation, but it can protect pieces of the system that actually make it valuable. If value is portable and transferrable to other ecosystems (like in crypto), people can move from one system to another with minimal economic loss, but there might be other aspects that might be worth fighting for or maintaining. For one, there's little reason to stick with a given system if you can just sell your portion for another form of value at a fair market price.

Goods like property rights, reputations, or even a corporation's labor supply all have limitations on the ability to transfer to other ecosystems. Although it's possible for someone to sell their house and leave their neighborhood and country, it will never be a seamless financial decision, as it's quite costly at a personal level. It takes time, and the relationships within your given community cannot be sold if you "opt out" of your country. For this reason, you're way more likely to see people fight and even have revolutions over their ability to influence change and affect their government.

Video games and social networks clearly don't have this level of social consideration, but when trying to create something that has lasting purpose, there is a choice regarding the number of frictions in leaving or entering the system. If you want people to hang around with others, you need to design it very carefully. If you can get paid to leave, you just might.

end [blockchain] games

Like all human interactions, many of our digital systems are non-economic and inefficient. There are factors outside of purely financial profit that could be considered frictions in pure market transactions. From a social perspective, these frictions serve an important purpose.

Minimizing exit, building relationships, and creating enjoyable experiences can all be seen as desirable, yet inefficient from a market standpoint. Cryptocurrency and financialized markets often do the exact opposite. They push for opt-in/opt-out systems where hyper-liquidity allows those who want to leave to do so at market value at the push of a button. They would be better served focusing on how to enable connection within the system rather than enabling exit from it.

Users can constrain these platforms by moving away from the systems once they financialize, but it might be too late at that point. Allowing exit from social systems via financialization is destabilizing. Incentivizing repeated hype cycles of these systems so early investors can grift later entrants is not sustainable.

One lesson that resonates throughout crypto is that you get

what you pay for, but nothing else. If you give inflationary rewards to hash power, you get lots and lots of hash power. What you don't get is an equitable distribution, fair access to a network, or a lasting purpose. The market delivers the cheapest version of your stated goal. If you want "players" or "users," you'll get bots or farming operations. If you want capital, you get the cheapest and most toxic on the planet.

The market optimizes for the metric, selling every desire you have that's not bolted onto your purchasing decision.

For those that do run blockchain games or social, this isn't to say that no web3-based social system will ever exist. It's just that transportable value will never be the selling point. There's little reason to think that a game or social platform will be successful because of the decentralization of the system. People want to be part of a community. Financialization does nothing to enable this.

~

"Oh my God, I totally want to live here!" she said as she jumped out of the car.

The couple met eyes for a few seconds. They both smiled and then looked at the house as they waited for their agent to get out of the car.

"It's a four-bed, three-bath, and this is right where you guys wanted," he said as he made his way up to the lockbox.

It was a nice place.

Beautiful brick row home, a block away from the park, a gorgeous backyard, and a screened-in porch to top it off. It wasn't new, but for Bennet and his wife, it was a dream that they could get something like this.

"How much is this one?" Bennet asked, slightly wincing to prepare for the answer.

"They're asking 900, but we'll probably have to go well over it and put in an offer ASAP if you want it," he replied.

The couple finally walked through the door to the spacious open floor plan. Bennet's wife was instantly across the room and talking to herself as she took in the feeling of the staged home.

"Not bad, oh, and I'm Bennet, by the way; don't think we've actually met."

"Of course... Jake," the realtor said as the two men shook hands. The two men continued walking through the house to keep up with the person clearly making the final decision.

"Your wife tells me you do crypto? So do you like mine bitcoin or something?" Jake asked.

"Ah, nah," laughed Bennet as he quietly looked for a way out of the conversation. "I actually run a different token."

"No way man, which one is it?" he said, now clearly invested in the conversation. "Something I can buy?"

"Um, yeah, DOJ, or Dojima, we're on the big exchanges. It's

kind of high now, so no promises if you know what I mean," said Bennet sheepishly.

"Nah, for sure man," said the real estate agent as he typed the information into his phone.

These kinds of conversations always left Bennet uncomfortable. The last thing he wanted was to be responsible for this guy losing money on his token. He had no idea which way the price would go and basically hated even trying to guess.

People would always look at the market cap of some 8-9 figure amount and assume he was loaded or a scammer. Or both.

He usually told people he was in finance if he could.

Jake paused for a second while he typed it into his phone. "Crazy man. I wish I had gotten into it. I heard about it a few years ago back on some podcast and thought it sounded awesome. Should have pulled the trigger... you must be crazy rich from getting in so early if you started one."

The realtor looked at Bennet curiously for clues.

Bennet sighed. He definitely wasn't out of the rat race in his mind. It was going well, but the company had honestly only had runway for the past year. He finally was able to sell a portion of his vested tokens, and his wife was amazed when they paid off her student loans, and he told her they could afford a house. He wasn't really a Bitcoin millionaire or living like a king yet, but he felt lucky to say the least.

He decided to ignore Jake's non-question.

The two men were standing on the back porch when Bennet's wife came running up to them smiling from ear to ear.

"So can we get it?!"

9

LIQUIDITY AND EXIT

"He grasped the fact that what appeared primarily an economic problem was essentially a social one. In economic terms the worker was certainly exploited: he did not get in exchange that which was his due. But more important though this was, it was far from all. In spite of exploitation, he might have been financially better off than before. But a principle quite unfavorable to individual and general happiness was wreaking havoc with his social environment, his neighborhood, his standing in the community, his craft; in a word, with those relationships to nature and man in which his economic existence was formerly embedded. The Industrial Revolution was causing a social dislocation of stupendous proportions, and the problem of poverty was merely the economic aspect of this event."
Karl Polyani – The Great Transformation

Imagine an island deep in the Pacific.

The inhabitants call it Yani.

Only a few miles long and sparsely populated, the few hundred islanders who live there survive by farming and fishing peacefully from their huts. One day a young economics student from America comes along and sees the citizens wasting their God-given capitalist potential. Convinced that stronger property rights would unlock dormant economic gains, he proposes a novel solution of privatizing the ocean itself. The citizens would divide the surrounding waters into discrete, tradable parcels that they each could claim, develop, and monetize as they saw fit. If they do this, he tells them, everyone will optimize their piece of ocean to its greatest potential. They could take out loans for businesses using their new collateral. They could improve upon their slice of water without having to worry about other people encroaching on their rights. They could sell their portion when they're ready to retire.

"Amazing," they all say.

As soon as the system is implemented, each Yanian citizen feels great.

The following day, the student helps one old woman put her plot up for sale online. Almost instantly, rich investors from the US see it and quickly make an offer. The citizens are astonished! The market price for each piece of water is $50k per parcel. No one initially sells, but they feel wealthier than their wildest dreams as they're now $50,000 richer, and all they had to do was fill out some paperwork.

The new property rights also help to spark economic activity on the island. Several citizens take out loans secured by their parcels for hut improvements and new businesses. Others get loans for better fishing boats too, and the island gets richer as outside investment pours in, again all secured by their parcels.

It feels good.

But cracks begin to appear. Some citizens begin to feel like their parcels are being overfished. The bigger fishing boats bought by loans are letting some people catch more fish than others, so to protect their slice and keep the free riders at bay, each plot owner throws large nets and buoys in the water.

Unable to even fish from their huts now (most of the fish have gone thanks to the nets), many of the citizens start to sell their parcels and leave. After the first year, a few loans go bad and more people sell. Before the islanders can protest, half of Yani is owned by foreign investors.

The parcels at the end of the story are $200k. The island as a whole is richer than ever, and Yani is a popular tourist spot.

~

Whether something can be financialized is far from the only question.

Tokens enable access to digital markets for previously stagnant forms of value. Similar to the property rights in Yani, they create "wealth" by allowing the value to be traded on the market. Formalization and tradability are the first step, but they also need liquidity in order to be deemed valuable.

In every cryptocurrency community, there's a common cry heard from early token holders: "We need more liquidity!" "If only we could get on XYZ exchange and get some real volume behind our token," or the less subtle "We should hire market makers for the token."

Liquidity is hailed as a godsend in the crypto world. It's easy to make the asset digital and tradable; the hard part is having someone willing to buy it. The theory is that if your token has more liquidity, this liquidity will beget large buyers, which in turn will increase the price and allow early holders to profit. To them, more liquidity means a greater ability to buy and sell, and this is always a good thing.

As a result of this view, many crypto projects spend large portions of their early revenue and investments on things such as exchange listings or market makers, all in order to boost this metric.

In practice, this theory of liquidity has had serious consequences.

crypto's superpower

Liquidity is the ability to enter or exit a system.

The usual definition is that it's the availability of an asset, usually referring to the ability to buy or sell in large amounts without materially affecting the price.

If the price does move, the amount by which you move it up is called "slippage." Usually thought of as a bad thing, it's often seen as a fee for buying the asset. This view makes sense in terms of trading for a dollar.

If Bob issues a token that represents one dollar in his bank account, it would not make much sense to pay much over a dollar for it. That said, if I really want to sell something right now, and the only way to do so is to accept his tokenized dollars, I might be willing to take a haircut on their value.

This inability to access the asset at its actual price represents a liquidity issue. Bob can always deposit more dollars to tokenize, but if the reason someone needs to pay more is because he hasn't done so (and therefore the exchange is limited in supply), this is a lack of liquidity.

In these cases, subsidizing liquidity can make sense. Bob making sure his asset has enough liquidity is important for the usage of Bob's tokenized dollars. That said, this concept of limiting slippage doesn't necessarily work as well with illiquid, highly volatile assets or assets where the buying and selling of the asset can materially affect the price itself.

FOR MOST ASSETS, liquidity is a symptom of utility and usage, not necessarily a desired property. For assets that look like equity or ownership, artificially introducing efficient trading and liquidity can create negative effects on both the owners and the system itself.

Imagine a scenario where you have a naturally liquid asset like oil. There are countless people who need to buy it (gas stations, airlines, etc.) and quite a few people who need to sell it (e.g., Saudi Arabia, Shell, BP, etc.). The buyers need to buy it on a frequent basis, and the suppliers get more on a daily basis and want to sell as fast as they can.

This is natural liquidity.

Lots of buying, lots of selling.

For crypto, this is rarely the case. Especially early on in the development of a chain, there are often no natural buyers. Most crypto tokens seek to be used as some form of payment (for transactions) or collateral for securing the network. The sellers will be the validators or parties given the token at its creation, but the buyers are almost always just speculators. With no users or applications at the outset, it's amazing there's any liquidity. Shockingly, however, crypto has solved the problem of liquidity. Even for the most obscure of assets, these systems can almost always find ways to make the trade happen.

If we measure liquidity by orderbook depth (how much you can buy or sell without moving the price), crypto has lots and lots of liquidity. Relative to any measure of usage or utility, even altcoins (smaller non-Bitcoin assets) have enormous market depth. Millions and millions of dollars' worth of instantly available liquidity for assets whose use case is still dubious.

Other metrics of liquidity are just as promising.

Trading volumes are enormous. Bitcoin has tens of billions of dollars in daily volume (about 50% of the crypto market in total), and other cryptocurrencies too are quite high by this measure.

To illustrate a comparison of how liquid crypto is, the industry has made assets that often achieve more liquidity than

S&P 500 futures contracts (some of the most liquid financial contracts in the world)!

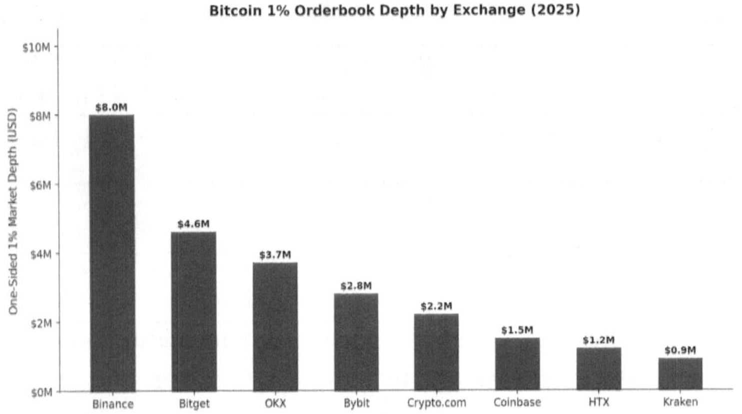

Bitcoin 1% Orderbook Depth by Exchange, 2025. Author's own figure.

Crypto's superpower is generating liquidity. [i]

The digital asset marketplaces that have developed for cryptocurrencies boast wildly liquid markets for completely speculative assets. The old days of having a hard time pricing an asset or finding a buyer are over. What would be normally bespoke financial assets traded in a peer-to-peer fashion (e.g., stocks of very early-stage companies or speculative commodities), crypto has enabled these same products all of the benefits of financialization, including hundreds of millions of dollars of global volume at very early stages.

If you have an asset that you turn into a token, crypto can make it liquid. The technology has enabled every asset to have a buyer and a seller.

As a result, many rush into this arrangement without questioning whether this form of financialization has actually been good for the systems that tap into it. Like many new technologies, just because companies can put the cart before the horse doesn't mean they necessarily should.

If liquidity means insiders can exit more easily, is this something that should be enabled?

When should something be illiquid?

external effects

Some will argue that liquidity is freedom. It allows anyone to enter or exit the system freely. The theory, especially in crypto, is that liquidity should be supported because everyone has a right to enter the system and also the right to leave with their portion of the value. Unfortunately, when it comes at the price of the system's sustainability and the quality of its interactions, it's not a freedom your system should offer.

Traditional venture capital is an example of illiquidity working well as an alignment mechanism. Whereas later-stage investors can buy or sell as they please, VCs who invest in early startups usually have multi-year lockup periods. It forces the investor to take a role as a committed partner in the company. This is seen in action too. They often help develop code, market the product, and assist in future rounds. The illiquidity incentivizes them to meaningfully help while their investment is locked. They don't have the right to exit. If they expect a return, they better help you grow.

And this makes sense. You want your investors tied to your system. If the going gets rough, you want to be able to call them for advice, have them give you some ideas, and use their connections to your advantage. The last thing you want is to allow them to leave. You wouldn't even call them if they could.

With crypto, much of that notion has gone out the window.

Early investors can buy into a project, and they can cash out anytime. There's no long-term incentive, and this actually hurts founders. A fast-moving software company is still figuring out what works. They need constant iterations and an ability to change paths with as little friction as possible. The last thing they need is to worry about the opinion of every investor or equity holder they may have.

Liquidity and the ability for anyone to buy in or sell out of your company have pushed founders right back to the limited transparency and short-term moves that plague public corporations. The founders are turned into mini-Fortune 500 CEOs, constantly managing perceptions to support their price. They focus on constant narrative pivots and profitable exits rather than actual product development.

internal effects

There are larger societal and systemic effects to providing liquidity to an entire class of assets. When it comes to crypto specifically, it's a predictable story as to what would happen if all early-stage companies were granted instant liquidity.

Why build a great product when you can just announce a great product?

One would assume that founders would require a longer lockup, but the market has proven so eager for speculative investments that parties can fake their way to success with enough capital.

Crypto has learned the hard way that liquidity unequally benefits informed traders and insiders.

As with most markets, liquidity implies open information regarding the traded assets. The theory of markets is such that all relative information should be priced in. This is fundamental to markets working. If someone knows a future action will move the price up, they buy up the asset as soon as possible. The valuation of a company represents all available information. If AI is worth $20 trillion, it's because the relevant information says this is a risk-adjusted, proper valuation. If a coin is selling for millions and you don't know why, it clearly means there's information you don't know. The market always represents the sum of all information available.

Even if you assume this to be correct, not all parties are privy to the same information. This is true with every market, but it's dangerous when it systematically favors one group over another.

With any early-stage startup, asymmetric information is everywhere. The founder knows the runway and the true runway. They know the customers on the press release and the ones who will actually cross the line to being profitable. They know the roadmap and the things that will actually pan out.

When companies have real customers and revenue, things start to make more sense to those on the outside. It even just gets harder to keep all of those with asymmetric information in line (quite a few more employees). Especially at the level of a large publicly traded company, at least in the US, there are disclosure requirements which force almost everything to be public information. The rest is literally illegal to trade against (insider trading). For startups and grey areas like crypto, this just isn't true. Many individuals know market-moving information well before the public, and in the past, they just couldn't trade on it.

In the case of our islanders on Yani, the same asymmetric information would create winners and losers. An owner who knows a resort is coming might buy up his neighbor's parcels. If someone does research and finds that all the fish are gone, they might sell their property before everyone realizes and the prices collapse. The liquidity gives them this exit. The question is whether the ability to buy or sell large quantities makes the community better.

If you don't have the inside information, you should fear liquidity.

There are countless stories in crypto of founders that buy

before they make a large partnership announcement. If there's enough liquidity, they can buy tokens before the announcement and dump them on the market afterwards. This hurts all of those new buyers who now have to buy in at a higher price.

Conversely, there are also parties who sell on insider information. Whereas the buying scenario is just unfair, the more important action for the current holders is the selling of the asset.

Imagine that a project gets a letter from the SEC or finds a bug in their code. If there's enough liquidity, the founders can dump their tokens on the market. The VCs who they call for advice get informed before the community and might even dump their shares before the team. It's a dangerous game, and retail participants are completely out of control.

As a general rule, the more information asymmetry there is, the less liquidity you want. Retail, in either case, should want very little liquidity without insider trading protections.

If parties are trading assets like dollars and euros, there aren't many secrets that materially move the price. For most equity, however, and especially at early stages with pseudonymous founders, the complete opposite is true.

finding the purpose

The liquidity necessary for an asset should be measured against the goal of the system itself.

For an example, let's use community currencies.

Imagine a city decides to start a "town dollar." They get together and give some amount of currency to every citizen, and all of the local businesses agree to accept the new currency as dollars. It works because rather than spending money at the Walmart and Starbucks that don't take our currency, everyone spends it at the local option, who will then pay their employees with it or spend it on something else local. The value of the interaction is kept inside the system. The 2% Walmart profit rate isn't siphoned over to Arkansas and the Walton family with every transaction, and best of all, it can give the local businesses a fighting chance.

Now for the million-dollar question: would an increase in financial liquidity be a good thing for the town dollar?

It depends.

Of course, you need a base level of financialization (people need to be able to spend it, trade it, and get more of it), but you don't want too much. No one should be able to exit too easily.

If the mom-and-pop store that accepts your currency doesn't pay their employees with the currency but instead sells it for dollars, the cycle stops. If Walmart can accept the currency but then instantly sell it at no loss for dollars, the entire system failed to shift any real activity in the economy.

It's counterintuitive at first, but the liquidity that makes it easier for people to trade for dollars actually destroys the entire purpose of the currency.

removal of value

Many of the economic challenges that come with introducing market systems stem from value exiting too easily. Liquidity is just the measure of how much can go at once.

It's hard to point the finger at crypto too much, but it's a prime example of systems affected by the market failure of financialization. The world as a whole is in a transition from economically closed systems to those open on a global scale. Increasingly, this means moving to market designs that allow people to move in and out of systems seamlessly. It has its benefits, but this efficiency-increasing digital revolution runs into problems when the structures and people around these interactions have long assumed them to be closed.

Thinking back to our community currency, small towns and

many human interactions were built on reciprocal and trust-based relationships; they assumed the value stayed within the system.

The lack of efficiency was fine and even part of the charm.

If everything is closed, exact (or efficient) pricing isn't as important. If you give money to the farmer to buy food for your store, it's ok if he can't make change. It doesn't have to be perfect, because he's probably going to spend it at your store. If one of them overpays slightly at the restaurant, it's a complex relationship, and he too will eventually give it back to the system.

In fact, the situation in some ways looks more like a family than an adversarial trade relationship. If the one farmer in your town is in financial trouble or has a rough patch, everyone knows they'd contribute to help out. The community needs him and benefits from him being part of it. The entire system is collaborative, and the precision of the market interaction for each transaction is less important.

Compare this to the example of a big box store that comes into your town and sends 2% of all sales to investors and 50% to foreign producers. This is extraction from your system, and unless you're getting it back in some way (e.g., remote corporate jobs), overpaying is purely a negative.

For this reason, liquidity often needs to be given in context to where it's going. Market systems work best when they are explicit about the area in which they are closed, and they need to be very conscious about it. If someone can extract value from one social layer and exit it to another (e.g., tax evasion or arbitrage opportunities by non-participants), it can often lead to undesired long-run equilibria, often a slow extraction from one system and disdain for the arrangement as a whole.

stand-alone markets

It's great when buyers can buy and sellers can sell out at a fair price. It's great when an asset is readily available to buy in many places so any user can access it easily. If a new blockchain wants people to be able to buy their token to stake it or use it to pay for a transaction, liquidity is super important.

But there's a big difference between buying ten dollars' worth of a coin to use the blockchain versus buying $10 million to capture a controlling stake. There's also a big difference between buying an investment and buying ownership in a community. Sometimes it's hard to separate the two, but as a general rule, if the ownership change resulting from a buy or sell materially affects the value, then the price should reflect that.

If you buy 1% of the token, it stands to reason you should move the price. If an asset represents a new ecosystem, the liquidity should represent that. If the founder of a protocol sells his shares, this might make the remaining shares considerably less valuable. Liquidity should represent that.

Modern digital finance can summon obscene amounts of liquidity. The question is whether we want to.

As on the island of Yani or with the local town dollar, financialization is not a panacea. Lack of liquidity on something can seem backward, but forcing trust and ties to a system can be beneficial. There are social consequences that make financialization, liquidity, and exit anything but fair. If you want engaged owners, citizens, or users that are tied to the community, protections need to be put in place to ensure that.

~

Standing in front of a banner of sponsors, Bennet looked down as the mic was being attached to his shirt collar.

The host of the podcast, clearly a television anchor in his pre-crypto life, went down the agenda. "So, we'll just ask you a few questions. Answer the best you can, and we'll upload it to a video montage with other founders. Sound good?"

Bennet nodded, slightly blinded from the immense lighting setup of the small stage.

The cameraman gave a thumbs up.

"Bennet, great to see you. How's your UpCon Hong Kong going?"

Bennet put on a smile: "It's amazing. First time in Asia, and it's been everything I'd expected. The investors and the builders are everywhere. It really gets me excited for the space."

"So, tell us, how does one get to be the CEO of a defi chain?"

He tried to remember to look at the host and not the camera.

"Yeah, so I came from the TradFi world. Maybe a little typical, but after Wharton, I started working as a derivatives trader at a macro fund. It was a phenomenal job, to be honest. I got exposure to a lot of different asset classes and trading instruments, but it was actually my personal interest in crypto that allowed me to break into the space. Some traders at my firm were making a killing on some early ICOs, and I was quick to get involved. I just took it a bit further and started coding so I could experiment with what the technology could do for finance. A little later, I left my job to get Dojima going."

He felt like that was good, but he put his hands into his pockets to avoid looking like a politician. The host gave him a nod of assurance.

"So, what makes Dojima different from the others?"

"It's decentralization and our community," said Bennet matter-of-factly. "True censorship resistance isn't something we take lightly. We know that if we succeed, we're going to be the backbone of global finance, and everyone who uses our system needs to know that they'll always have access to their funds and the ability to trade, and that's what we do."

"Nice, and I have to ask, you guys are seeing a lot of Asian volume on your decentralized stocks protocol. How did you finally make Tesla on the blockchain a reality?"

Bennet appreciated that the host had clearly done his homework.

"Well, to be fair, you're not trading the stock directly. A lot of our competitors were trying to put actual stocks on the blockchain, but we made the bet that users don't actually want that. They just want access to the price movement. We have a proprietary system for pricing and settlement, and it enables us to safely and in a completely decentralized manner allow anyone to take a position, long or short, in any stock they want."

"Two more questions: Can you explain the new initiative to get liquidity on the system? I think a lot of people are coming to your system to get rewards."

"Sure," said Bennet calmly. "It was just part of our program to create a sustainable token ecosystem. The liquidity incentives help us find the product that users want, and the staking incentives work to make it a flywheel where the interest in the tokenized trading directly contributes to the security and longevity of the system."

He felt proud of himself after that technical description. At this point he was almost a machine.

"Now before we go any further, where can I get this shirt you're wearing? If you're not watching, this is designer-quality swag for your team I've seen all over the venue, and I just have to say, I love it."

10

PREDICTION MARKETS

"Society is something that precedes the individual. Anyone who either cannot lead the common life or is so self-sufficient as not to need to, and therefore does not partake of society, is either a beast or a god."

–Aristotle, Politics

In 2025, there was a popular prediction market on whether Taylor Swift and Travis Kelce would get engaged. It had several hundred thousand dollars in volume of mainly people saying 'no.' At the end of August, the odds stood at 24% that they would be married by year's end.

The idea of betting on someone's engagement might come as a shock, but pop culture markets of the sort were relatively common, so no one thought much of the activity. Compared to some other celebrity categories, it was relatively thinly traded.

To everyone's surprise, however, one afternoon someone placed a huge bet on the platform, roughly doubling the odds.[i] The gambler seemed rather certain that Taylor and Travis might actually be getting engaged this year. The next day, Taylor posted an announcement on Instagram, and the market was closed. Our speculator was correct.

No one knows who it was, although some rumors spread that it was likely her guitarist.[ii] Whoever it was, they clearly knew something the rest of us didn't and revealed the truth to the world. When the market was settled, hundreds of people lost money, and a handful of people made money by being either the speculator himself or the first to arbitrage the news and drive the market to certainty.

Proponents of prediction markets were ecstatic. The system beat the news stations and social media accounts to the story. Our hero revealed valuable information a whole twelve hours before her announcement. [iii]

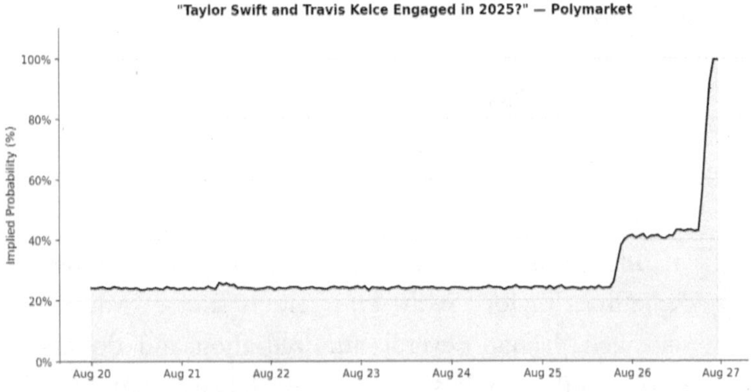

Polymarket prediction market: "Taylor Swift and Travis Kelce Engaged in 2025?". Author's own figure.

FINANCIAL MARKETS REVEALING TRUTH. This is the goal of prediction markets.

Commonly called betting markets or information markets,

prediction markets are more than gambling in that they have a greater aim: to reveal information. In this sense, they're pitched as an alternative to surveys and polling. Similar to other financial tools that price risk or determine an equilibrium, the self-interest of individuals gambling should theoretically lead to information being revealed (like if Taylor will be engaged).

Just as bonds can be used to calculate the odds that a creditor will default or insurance premiums can be used to find the likelihood of an unfortunate event happening, these market-style pricing mechanisms can be used for all questions or statements. Now instead of just being restricted to financial markets, the promise is that this targeted application of gambling can lead to more knowledgeable and honest systems across the board.

why not just surveys?

Signaling, Selection, and Noise.

Humans have long held a bias toward wishful thinking. When asked what might happen, we often choose preference over probability. When asked via survey or in person, we rarely give an honest or analytical assessment but rather our subconscious emotional investment in a particular outcome.

This same bias is true for reporting and even official statements as well. Often news organizations, social media personalities, or even government officials will let the world know what they want you to think is true, rather than what is.

Anytime a signal could be derived from the statement, there's a potential for this kind of bias, and this is where prediction markets can assist.

We all have experienced this bias firsthand. If you've ever asked someone, "Who do you think will win the football game?" you know the respondent will often answer who they would like to win, even if they deep down may not give much hope to their hometown favorite. Asking them to bet on who they think will win removes this bias if they place more care in losing the economic collateral than in signaling with their prediction. Anonymous surveys and proper question design can help alleviate this problem too, but having skin in the game in terms of economic collateral can move the needle even further.

Surveys are also limited by who is being asked the questions.

An optional survey tends to get only people who respond to surveys. A random sample depends on the population you're sampling from and their expertise regarding the question. A survey of uninformed participants provides no information. If you were to ask ten people who they think will win a football game, but nine of these individuals have no idea how football even works, the survey is of little use if the uninformed answers are not distinguishable from the one sampled who's very knowledgeable on the subject. An open betting market not only allows self-selection in a good way but also incentivizes any expert with an opinion to come and state their prediction with magnitude (i.e., they can bet more).

Lastly, there's often not just one survey.

As is the case in presidential elections, there are often tens or even hundreds of surveys and polls with various methodologies and countless special interests funding specific results. Unless you're an expert in analyzing their design and willing to put in the time and effort to understand each claim, competing surveys can be confusing if there's no way to credibly sort through the noise.

take off

The idea of using prediction markets isn't necessarily new, but since the information is an externality of the betting, the actual deployment of these systems has been highly regulated due to various factors, namely historical thoughts on gambling and consumer protection laws. For these reasons, as well as the overall blanket ban in the US and limitations in other parts of the world, prediction markets have stayed relatively obscure over the past few decades, despite interest.

But that's beginning to change thanks to our favorite regulation-murkying technology, which is crypto.

People recognized that crypto could be used for prediction markets almost from the outset. As early as 2014, developers were trying to use the permissionless nature of Bitcoin to create prediction markets and thus, information that could not be censored.[iv] Early on, the designs were constrained due to the limitations of the technology, but with Ethereum and the rise of smart contract platforms, the systems began to take off. Early iterations around 2015 began showing working products (e.g., Augur[v]). Unfortunately, due to a mix of the nascency of crypto and some early regulatory concerns, they failed to really achieve more than academic or hobbyist attention.

Some in the field are even pushing to use prediction markets to make and automate political and financial decisions. Dubbed "futarchy,"[vi] this subset of prediction markets creates systems where protocols or companies use these markets to actually make corporate or governance decisions.

As an example, imagine a company needs to decide where to deploy funds. One option is to use the money to build a new headquarters. The other is to hire a new research team that will attempt to find the company's next big product. Rather than convening experts or the board, futarchy would have them create several prediction markets where parties would place bets on what the effect of each action would be on the future share price (e.g., "Will the stock price rise by 10% over the next year following this action?"). The proposal that the markets reveal to be the most beneficial to the share price gets the funding.

Futarchy aims to remove all governance centralization and even remove the need for management. All decisions could be handled by the market. It's a neat concept but so far limited in practice for many reasons.

For the most part, information markets remained a theoretical exercise in crypto for many years. But then in 2024, a new crop of prediction markets had arrived and were ready to capitalize on the moment that was the presidential election. [vii]

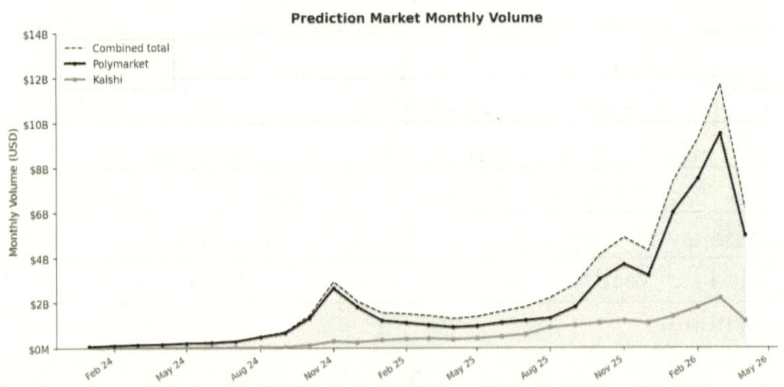

Prediction Market Monthly Volume, January 2024–April 2026. Combined Polymarket and Kalshi taker notional volume. April 2026 is a partial-month estimate. Author's own figure.

THE POPULARITY of these platforms skyrocketed. Billions of dollars were being bet on these platforms to try and predict the outcome of the 2024 election. By the day of the election, where most polls were completely neutral, the prediction markets strongly favored Trump. [viii]

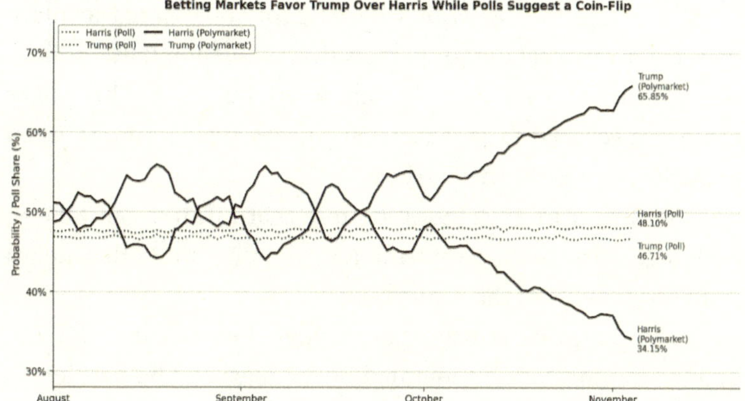

Betting markets versus polling averages for the 2024 U.S. presidential election, Author's own figure.

PREDICTION MARKETS GOT IT RIGHT, and they got it right faster than traditional polls. As soon as the election was over, innovators and existing companies raised significant capital on the promise of expanding prediction markets to all events and even decisions in our lives. Between prediction markets and AI, the vision was to provide certainty (or slightly more honesty) to everything from news stories and social media to corporate governance.

But of course, as far as the election is concerned, correctly identifying the outcome of a coin toss could be just luck. Does the theory actually make sense?

game theoretical outcome of prediction markets

The informational quality of a given prediction market is determined by the intersection of two questions:

- Who do we want to participate in the market?
- Who will participate in the market?

When it comes to which participants one wants to place bets, from a market design perspective, prediction markets definitely don't want insiders that know the outcome. This might come as a surprise to some, as many times, insiders are claimed to be the purpose of the market.

"Prediction markets are a bounty that insiders can take in exchange for revealing information" is something often heard from proponents of these markets.

To be fair, this is true. The insider does wonders for the quality of information generated by the market. The problem is that they're not being used this way. None of the gamblers who lost money on Taylor Swift's engagement looked at their bet as a bounty to an insider. No one who bets on a sports game is trying to throw the game by placing bets as bribes. If this was the goal, there are more capital-efficient ways to do this (why give gamblers anything?).

If the participants are gamblers, over any repeated game scenario, insiders destroy the longevity of the platform. If all other participants are certain to lose, people tend to not play. In traditional finance this is insider trading, and it's illegal for a reason; even if you're a good trader, you'll always lose money to them. Why would anyone ever bet on a sports game if they know the athletes or refs are betting?

So, assuming that insiders are not desired, the participant class wanted for actual information is informed parties. Uninformed parties are just opinionated noise. Prediction markets whose goal is information should want people who have a reason to believe something enough to put money on it, preferably in a fashion that indicates their certainty.

But here's the catch: do those informed participants disagree on things, and are they risk-takers? If they are not risk takers, they're not playing this game. Most of these prediction markets we care about aren't necessarily repeat games, so if they think they might lose their money, they won't take part.

These systems need them to disagree so they can bet against each other too. If a market has 30% of experts who believe one thing and 70% who believe the alternative, now it can function, and a prediction market can make sense. But if the experts all agree, the market needs uninformed gamblers to take the other side, or there's no one betting.

And this is where most prediction markets spend their time: getting uninformed gamblers to lose money to the informed participants. Just like sports betting platforms, prediction markets need the guys throwing money at their home team so you can have some fodder for people actually crunching the numbers (in many cases, the house). Only then can you get a market that reveals an informative price.

But as for who actually participates, most of the time, you don't get anyone.

This is the problem for prediction markets at the moment. Game theory suggests that uninformed participants should abstain entirely. Sports betting, or the closely related presidential election betting, is driven by emotional factors and beliefs that push retail participation beyond just rational predictions for rewards. Unfortunately for most information markets, they just don't draw the same amount of excitement or investment.

the problems of financializing information

Besides the game theory of participants, financializing information has several other problems that limit its effectiveness.

The first is that of capital efficiency.

Money is pricey. Money for a long time is even pricier.

One huge issue for prediction markets is that capital needs to be locked in them. If users want to know who will win a football game tomorrow, this isn't a problem, but if they want to know whether the US is likely to end relations with a foreign country over the next 3 years, it becomes much more difficult thanks to the significant carrying costs for locking up that capital.

For futarchy, this means that any decision a company wants to make will require participants locking up capital for the duration of that decision. If you want to bet whether the stock price will rise in the next 5 years, this is much more costly than asking whether it will rise by EOY.

We see this in practice too. Most of the volume, and thus information, in prediction markets comes in the weeks or days before settlement. If you look at the presidential election market, the volumes took off in October. In fact, the markets even had Harris leading for part of September. What this means is that prediction markets reveal information, but only slightly before they have to. For the Taylor Swift engagement, the information was revealed twelve hours before; for the presidential election, just several weeks.

Playing out the game theory, it forces anyone using these markets for information to be focused on the short term. Futarchy would mean that corporate decisions are systematically skewed toward short-term endeavors.

market feedback and real-world effects

Another example of a questionable success for prediction markets came in early 2026. At the time, growing tensions between the United States and Venezuela led to a market tied to whether the US would take actions regarding the Maduro regime.[ix]

Despite months of activity, the odds were relatively low going into the new year. Trump had been threatening action, but the odds of Maduro being out by the end of the year were still around 6%.

Then, all of a sudden, one gambler placed massive bets

totaling over $30 million on the prediction market. After the whale's bets, the odds shot up to over 50%.

The next day, news broke that the US captured President Maduro. [x]

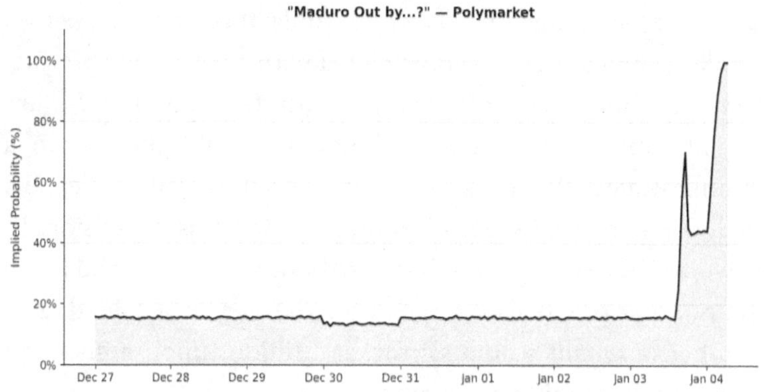

Polymarket prediction market: "Maduro Out by...?" Settled January 31, 2026. Author's own figure.

If prediction markets are large enough, they can affect the outcome of the underlying event. In sports, players can throw the game. In finance, spot markets are often manipulated to affect derivatives pricing.[xi] This is specifically illegal, and yet it still happens. If a market is used to make decisions for a corporation (as with futarchy), it's inevitable that someone will try and game the market. These are real risks, and they shouldn't be discounted, but there are risks that come even from just revealing information.

It's a clear conflict of interest, and even worse, it can have serious consequences for national security and the effectiveness of the government. Leaking military actions before they happen, betting on Federal Reserve decisions, and betting on the actions of one's own vote can all have real-world repercussions.

fake news and information quality

Goodhart's law states, "*When a measure becomes a target, it ceases to be a good measure.*" The rule is the perfect summary of both financialization and information markets.

To explain, imagine a trader finds that the number of emails a company sends is correlated with a company's productivity and thus their profit the next quarter. Assuming he has the information for each company, he could then write a trading strategy that buys stocks based on the number of emails each company sends. This is a metric informing a decision. If, however, those companies find out about the strategy and make the number of emails a target, now the metric is ruined. The number of emails would skyrocket. Millions upon millions of emails with meaningless content.

It ceases to be a good measure.

This is the problem with financialized information. These systems are not revealing truth, solely perceived or predicted information. If gamblers want to make money trading information, they don't need to change the outcome, only the perception of the odds of the outcome. If they signal that something will happen, trade the move higher, and get out (like you can in current systems), then all the system does is fuel even more misinformation. If a celebrity can announce a presidential run only to cancel a day later after betting on themselves to lose, does this help our current state of information overload? If fake partnership announcements and business deals move the needle, how do we verify news and prevent this kind of exploit?

Misinformation is bad now, but fueling it with financialization and pushing it onto every question in society is borderline insanity.

broader questions

There are endless websites, research studies, datasets, posts, and commentary for any question or even thought. We can check hundreds of different forums, government websites, news reports, or LLMs telling us exactly what the truth is.

And yet we still disagree on basic facts.

There's no need to place blame or develop some conspiracy behind it. Simply put, there was demand for information, and the market delivered.

We're drowning in "information."

This is where prediction markets want to help. We need some way to parse through the research and noise, and it's a noble goal to try and find solutions.

Unfortunately, prediction markets, at least in their current form, look to do more harm than good. In a nation with more insider manipulation than ever and gambling addictions at an all-time high, we need to be cautious if we claim prediction markets are the answer.

Yes, they allowed us to learn about Taylor Swift's engagement 12 hours before her official statement. From a societal perspective, however, is knowing whether Taylor Swift gets engaged a good thing worthy of having gamblers lose money? Is there any real information even released, or is it just front-running in a different medium? It's not clear that expanding these betting markets to financialize more information leads to a better system.

I recently read a story about how an insurance fraud case was solved by a dash cam. The ever-recording video camera on the front of the car showed a woman driving in an almost comical fashion (were it not so dangerous), all so she could get into a minor accident to file a claim. The news agency then explained how the prevalence of doorbell cameras and dashcam footage made it much harder for criminals to cheat the system. It even then elaborated on how AI was being used in many cases to pore over this footage to identify patterns of crime when they occurred (e.g., call the police if a masked man walks up to your door to steal packages).

This sounds great. Until you realize that the consumers are paying to install a panopticon.

Prediction markets fail theoretically because the incentives risk fueling misinformation and short-termism if adopted more broadly.

Prediction markets fail morally because in practice they depend on continuously getting new gamblers to lose money so insiders can extract value in exchange for information. This structure is nearly identical to gambling or even Ponzi schemes. Even if they can help to cut through some of the noise, sometimes the cure is worse than the disease.

Markets are all about externalities.

The argument that it worked for the presidential market is not a good one.

As soon as the metric becomes the target, information will be gamed, and the results will show even less success. Until substantial structural and regulatory changes are made, societies can do better than introducing this obviously lacking structure into a position of prominence.

~

"Hello again, sir. Long time no see," said Marco over the Zoom call.

"Too long man," said Bennet as he adjusted his camera. "Looking huge by the way, congrats."

"I'm doing what I can," said Marco with a smile. "Thank you for taking the call. I won't waste your time, but I wanted to see if you'd be one of the delegates for our DAO."

Bennet paused. A little taken aback, he didn't know what to say.

He knew people who had these roles, but he had always been relatively critical of the "professional delegate." He noted his personal position shifting. Apparently, he realized he really just wanted to be invited to the club.

"Hell yeah, man," he said after a chuckle. "What's involved, and, sorry for asking, but can you go over the project again? I just want to make sure there's no conflicts with Dojima."

"No, shouldn't be. The job is pretty easy, to be honest. We'll be running the DAO mainly through our forum and Discord. You can come and introduce yourself and tell people they'll be able to delegate their tokens to you. We will delegate some to you from the foundation too. We need to split things up for the optics, but you'll get a portion of the inflationary rewards for your work," he said. "Just stay up-to-date on the project and vote on the upgrade and grant proposals."

Bennet knew Marco only from the few conferences they'd traveled to together. He was smart and hadn't been involved in any scams. To Bennet, this made him one of the good guys in the space.

Maybe it was too low of a bar, but most people he met in the space didn't really clear it.

A girl in a bikini walked behind Marco as he leaned back in his wicker chair. It looked like he was somewhere warm and far from the East Coast where Bennet was.

"This is the defi yield fund idea, right?" Bennet replied.

"Yes. We got a grant from several chains to launch it, and so we'll be launching the points program on testnet next week. And the official pitch is a 'tokenized strategy vault,'" he said bluntly.

Bennet knew enough to know his lawyer told him to use that phrase.

Marco continued, "We'll be using the DAO to find the best yield across any chain."

Bennet wasn't quite sure on the technicals or how a cross-chain operation of this sort could in any way be decentralized. He was also pretty sure that Marco knew this as well, but he couldn't deny that this was where the market was moving.

"Yeah, I think I'm on board," he said, not wanting to talk himself out of a good thing. "Happy to help."

Dojima's runway was starting to dry up, and it looked like they were in the early innings of a new bear market. A little bit of diversification in what he did would be good for Dojima. The company needed more exposure in this new "tokenized fund" based world, and if he could figure out a strategy, Dojima should pay him more for doing this. And even on the surface, if he could push some of these funds to Dojima, it would be a win-win.

At the very least, he convinced himself that was the case.

"And how's the family?" asked Marco. "I think you got a house right, is this it?" he said, looking at Bennet's background.

"It's going great. But yeah, I mean no, this is just the coworking space we rent for the team. Just me today, so I'm nice and spread out. But you gotta come out this way one of these days too."

"I think I might have to. This travelling is starting to wear on me. I feel like I am burning out from working so much, if you know what I mean."

"Of course," said Bennet.

11

FREEDOM AND MARKETS

"The direction of technology, its inequality implications, and the extent to which productivity gains are shared between capital and labor are not inescapable givens; they are societal choices. Once we accept this reality, the case that society should let technology go wherever powerful corporations and a small group of people want, then do its best by trying to catch up with education, seems less compelling. Rather, technology should be steered in a direction that best uses a workforce's skills..."

Daron Acemoglu and Simon Johnson, Power and Progress

Hours after dusk, three men gathered at the edge of town. They had heard rumors of this working to the west in Arnold but had honestly thought the whole lot mad when they first found out.

The only reason they actually believed it might succeed was thanks to the pushback from the authorities.

The month prior, the government had put out flyers decrying the activity as a "heinous and ungodly destruction of private property," a great sign it pissed the right people off. Several "official" statements even hinted at militia being recruited to put an end to it, but the night here seemed quiet as usual.

Over the past decade, these men and countless other families had migrated to what most of the old folk described as "an unholy urban metropolis," with factories and roads bursting at its seams toward the coast and the mines. It seemed that almost every person they grew up with had been forced to migrate to this city for work over the past decade.

And this wasn't even an urban versus rural qualm.

They never had much in the way of luxury. Life was hard and dirty, but what they had was self-respect. A plot of land or home that went back for centuries, a trade that your father and his father before him performed with whatever pride a man would have in such a situation. Life meant farming, large extended families, and drinking some ale with your community. But now it was wages, hours, filthy factories, and so-called "homes" in the city.

The owners living like kings, and the laborers doing the work.

Once deemed quiet enough, the group lurked around the outskirts of the town to the back of the factory. The leader of the group going first, he used his axe to quietly pry open the locks on the wooden door. The other men kept watch, honestly surprised as to how easily the locked door opened. Once inside, they made their way to the large knitting frame, where they silently cut several of the wires and stole a few of the critical pieces. Familiar with the workings of the machine, the men knew this act would render the machine useless. The owners would be forced to take the next week or two replacing it at a hefty cost. Best of all, they hopefully wouldn't get caught, for as far as they knew, they hadn't woken a soul.

After the deed was done, the men snuck out the same door in the back. They made their way home to their wives, who would be heading to that same factory early the next morning.

Later that year, an act was passed declaring the breaking of a knitting frame punishable by death.

New technologies aren't always welcome.

Before the mid-1800s, textile manufacturing (the making and weaving of clothes and fabric) was "decentralized." Individual textile companies did not have facilities for making and processing the materials themselves, so they would largely contract out the work to individual artisans and workers who would produce the final goods. Urban factories had been tried before but proved too costly. The arrangement with rural labor, on the other hand, could take advantage of women and children who could work out of their home. Additionally, these workers were closer to the source of material (cotton or wool) and many times would even be the ones who worked the farms or raised the cattle. In this way, the setup provided for a large number of rural communities that then shipped their goods to coastal cities for trade and sale.

It wasn't until the adoption of the spinning jenny in the late 1700s that mechanized textile production began to take hold. With this and similar inventions, the individual contracts with home artisans started drying up. Workers were pushed toward urban areas to be employed in standardized factories where they manned a portion of a machine.

These "labor-reducing" machines drastically increased the productivity of each worker.

This sounds positive, but the adoption led to drastic changes for workers. No longer were they "artisans" or skilled tradesmen, but rather just cogs in a machine. Employment also changed from life on a working farm to a job in the factory.

Few of the workers would willingly accept this tradeoff. Unfortunately, as the contract work dried up, it became all but mandatory. Long hours for little pay in poor conditions was the new norm. For many men, women, and children alike, this was the Industrial Revolution.

For these reasons, the workers began what would be the first of many movements for their rights, but more specifically, the anti-technology movement of the Luddites. These workers went on strike and even destroyed the new machines in protest. It would ultimately prove futile, but for the next hundred years, they would fight against automation, markets, and "progress."[i,ii]

Spinning - Hand-colored engraving, c. 1830s.

Luddites smashing textile machinery, early 19th century.

Like the spinning jenny, blockchains and cryptocurrencies are technologies that promise automation and progress. They too seek to bring about efficiency while giving little consideration to the social tradeoffs being made. Even within the brief timeline of crypto, it's easy to see that these new markets have influenced the participants in and institutions surrounding their existence.

Early on, crypto created relatively few problems. The only people interacting with or adopting the system were the enthusiasts that had the free time. Similar to early internet users, the biggest downside for the first few iterations of any protocol is simply the time wasted by people who probably don't mind. As the system gains momentum, however, the real consequences begin to appear, be it to external industries, social structures, or even other cryptocurrencies themselves.

It's not for a lack of trying either. The crypto industry tried to distribute ownership and responsibilities to hopefully remove control and even alleviate some of the distributional inequities present in traditional structures.

Unfortunately, results failed to materialize. Early adopters did find speculative gains, but the story of the technology's purpose played out differently. Whereas ownership of the platform was supposed to lead to a greater commitment and ties to the system, financialization and the instantly available liquidity led to holders simply cashing out after getting in early. Whereas systems were supposed to be built to remove the need for banks, democratize finance, or create personal ownership of data, their own neutrality has led to them being passively captured.

And we've heard this before. This happened to web2, traditional finance, manufacturing, social media, and so many more. This trend toward extractive equilibriums can be thought of as nothing but a feature of markets. Each new layer of financialization and neutrality creates new abstractions to hide the exploitation.

Historians over the past century have written extensively that innovation and markets can cause disruptions to existing industry. Joseph Schumpeter was one of the most famous economists of the early 1900s. On the heels of the industrial revolution, he coined the term "creative destruction" to describe what was an obvious outcome of markets at the time. The term encompassed what he designated as capitalism's essential feature, the continuous life cycle of economies, the destruction of the old to make way for the new. [iii]

A classic example would be that of the car replacing the carriage and horse. A new technology can come along and put many out of work.

He argued that this process was literally the defining characteristic of free markets, the "perennial gale" that would sweep away established firms, technologies, and entire industries all through nothing but innovation and competition.

Everyone sees it happening, but it's far from what economists are taught at university. The classical theory of markets is that a new factory comes to town and hires the old workers at a better rate, improving the situation across the board. Unfortunately, it never works like that. Instead, the competition is increasingly global, and the new workers are rarely the same ones as before. As the new technology emerges, investors pour in capital, and incumbent industries are left to decay. The skills of the outdated industry become obsolete. Communities built around those industries are abandoned and hollowed out. The global economy as a whole grows more productive, and new opportunities emerge, but it's elsewhere, for someone else.

Schumpeter saw this clearly, but what he underestimated was the social cost of the "eventually." The workers in the old industry couldn't simply retrain. For the Luddites, the textile workers weren't trained or educated enough to be engineers or merchants. In the US, the Appalachian mining towns that collapsed with coal demand didn't seamlessly transform into silicon chip factories. The truck drivers don't really learn to code.

The creative part of the process happens quickly. A new business or industry is formed, and the benefits flow to early innovators and investors. The destruction, however, can be slower. It's widely dispersed across communities and different industries that lose their economic foundation. The market removes blame from any one party for the demise of the industry, but the outcome is the same.

Seeing as crypto and digital markets promise to greatly accelerate the drive towards efficiency, it's no wonder that many people feel economic and social uncertainty. What would Schumpeter say if both the new and old could be destroyed with such ease?

unseen outcomes

Western society has accepted the theory that innovation is worth it. Meritocracy and inventions lead to prosperity, so the losers in the bargain need to adapt. Markets too help this arrangement. Any financialized system has an amazing property in its ability to abstract ownership over a decision. Not choosing is itself a choice. The adoption of a new technology that predictably causes losses for one subset of society is also a choice.

Crypto, like many technologists, views the disruptive abilities of their innovations as a selling point for the adoption. The tools that remove middlemen or provide the ability to defy regulations can be useful. The investments that shatter an existing industry to produce a marginally better option for modern consumers.

Of course, these technologies can be beneficial. Crypto can

prevent an authoritarian government, lead to transparency in pricing, or even help maintain privacy for marginalized groups. It can be used as a check on the mob-rule mentality that is democracy or even a limit on the controls afforded to parties in power.

But the utility is far from automatic.

What the technology is ultimately used for is still up for debate. These same systems might lead to more entrenched power structures. They might exacerbate wealth inequality and harm local governments in a push toward global efficiency.

The sacrifice of social norms or collective agreement to aggregated self-interests shouldn't be made lightly.

The consequences of the transition to markets can be seen in many industries. Digital marketplaces add value for consumers by moving the experience from in-person or even from multiple websites all to one platform. If there are 10 different sellers for a certain item, the customer can go to one site, pick the cheapest or top-rated, and move on. The customer couldn't be happier, and the prices stay low thanks to lots of competition driving down the price for a standard good.

The consequence is that the UX improvements abstract details about where or how it's made. Now a factory in the US with high-paying jobs and concern for the environment is outcompeted by a sweatshop in a foreign country. The efficient market creates a bounty for exploiting the unchecked assumption. It behaves as an anonymous incentive to reduce quality or costly values regarding production in favor of the examined metric of price. No specific customer is neutral, but the market obfuscates each individual's ability to discern which pieces matter. When the outcome is finally realized, each person feels powerless to change the collective outcome.

Competition often presents to you a new option without ever telling you about the repercussions.

The market designed to secure a database incentivized only

the brute quantity of hash power as the metric. As a result, it very quickly found an undesirable distribution and cleanliness of that hash power. Predictably needed goods like custodians, block explorers, nodes, and staking services were left to the market to deliver, and now these providers extract and influence the system.

This is the market behaving as markets do.

The efficient market settles on the person with the lowest cost eventually performing the task. In the same way a factory in a world void of friction will move to find the cheapest labor, digital systems will do the same, just much faster.

This could be a good thing if the goal is to create wage or cost competitions on a global scale, but oftentimes labor and resources in these situations are coerced and exploited in ways most of Western society largely deems as antithetical to progress. Individuals do care about who they interact with and how products are made. Removing the frictions that force this information upon us is not necessarily an economic miracle.

can unstoppable be put back in the bag?

Everyone has seen the pitches of crypto and AI founders telling of the radical transformations coming to the economy. "This sector will be automated." "That entire industry will be handled by code."

VCs buy in.

Retail investors argue over whether it's real and a good investment or not.

Everyone constantly asks "when" or "if," but very few whether they should. Society seems incapable of asking whether the unrestrained technology is worth the exacerbated inequality and massive unemployment.

Efficiency is inevitable, so it must be embraced at all costs.

Progress is a net good, so don't slow it down.

And it's not that people don't realize the dangers in these new technologies. Everyone knows that AI and automation can hurt labor markets. Everyone knows that crypto and financialization can lead to gambling or be used for scams. But everyone, everywhere, seems so beholden to the narrative of progress that any discussion of addressing its wake is seen as a waste of resources.

The freedom to innovate trumps the freedom to live without it.

If people want to have a system without these innovations, can they even do this anymore?

There is a choice, but it's collective and it's even harder to coordinate an answer.

As a result, "allow everything" has become the default.

We are presented with the choice of unrestrained capitalism that destroys the environment, automates all decent jobs, addicts society to slop content on screens, and leads to Dickens-level inequality, or we're told we lose to China and decay permanently through slow growth.

But these aren't the only options.

The Luddites were not generically anti-technology. They were specifically protesting the machinery that circumvented labor agreements and undercut skilled workers' wages and lifestyles. Even if inevitable, slowing down and thinking through the decision of which pieces work can help everyone involved.

Some pieces of crypto are helping. There are efforts to use novel, codified structures to better regulate communities and companies. There are experiments taking place to create better organizations that aren't as captured as modern corporate or democratic arrangements.

It's both possible to like some parts of crypto and also be cautious of unrestrained financialization and neutrality. One can recognize that despite the negative externalities of cryptocurrency, there is freedom in being able to move value and create contracts in a way that certain actors can't stop. Whether it's buying drugs online, anonymously donating to a protest movement, or saving money from a hyperinflationary collapse, there is a benefit in limiting the tyranny of traditional governments or institutions.

But which pieces should be kept is not set in stone. There's still time to pick what parts of society are subject to these systems and which parts we leave to more traditional forms of human discretion.

~

[MESSAGE FROM THE TEAM, DOJIMA DISCORD]

The Dojima chain remains paused for mitigation of a confirmed exploit in the liquidation handler mechanism. On Friday, an encoding error led to an oracle mispricing. The attacker used this error to drain several of the thinly traded markets.

Just over $18M in funds were transferred to CDUSD and bridged out of the system. We have worked with partners on the target chain as well as issuers of the stablecoin to black-list the address and all movement by the hacker. We will be seeking a full return of the funds and a restarting of the chain according to timelines set by the team. We are working with law enforcement and feel confident in the full safety of user funds.

A fix has been identified and will be deployed to secure the system from the exploit going forward. A full post-mortem will be published, and the team will continue to stay available for questions and will provide more updates on the chain's restarting as soon as possible.

Thank you for your continued patience and support.

12

FEE MARKETS - MEV AND SKILL GAMES

"It used to be 'the economy' was the way we provide people with food, shelter and other necessities to keep them alive. Now we're all supposed to die for the sake of the economy. What is this thing they're talking about. Is it just code for rich people, class power, or what?

-David Graeber

In 2012, the Obama administration announced across-the-board increases in fuel efficiency standards for new American vehicles. These regulations stated that all car manufacturers would be required to achieve an average of 54.5 miles per gallon by 2025.

It represented a roughly 80% increase from the current standards. The requirements had the noble goals of both combatting climate change and reducing dependence on foreign oil.

Environmental groups were ecstatic.

According to the administration, Americans were going to benefit by saving thousands in fuel costs, the car manufacturers would win by becoming the global leaders in clean technology, and of course, the planet would benefit from reduced carbon emissions.

The policy was the neoliberal dream. Expertly designed market regulation would harness the market for socially beneficial outcomes, a perfect compromise balancing left and right.

But the footnotes matter.

The regulators wanted to keep manufacturers from simply abandoning truck production altogether, so they couldn't just put extreme requirements on large vehicles. As a result, they instead set different targets based on a vehicle's size, specifically giving larger vehicles lower efficiency requirements than smaller ones.

Automakers quickly realized they could more easily meet the requirements by just making the vehicles larger. As an example, a Honda Civic at 36 MPG might miss its target, while a Ford F-150 that got 24 MPG would easily exceed its standard. Since the latter was both easier to meet and more profitable to sell, car manufacturers changed their entire strategy.

Aided by cheap gasoline prices, they advertised and pushed ever larger trucks and SUVs on consumers. At the same time, they raised prices on smaller vehicles, with many manufacturers largely abandoning much of the small car market. Within five years, nearly every company was discontinuing pieces or the entirety of their sedan production.

By 2019, light trucks and SUVs were nearly 70% of new vehicle sales. This was up from 50% when the regulations were proposed.

Total fuel consumption barely budged despite massive investments in efficiency technology. Even worse, American roads were transformed into a landscape of ever larger cars and trucks.

Similar to carbon in the environment, crypto has its own negative externality it's trying to reduce: maximum extractable value, or MEV.

MEV is the profit validators can extract by controlling the order of transactions.

To explain, the validator (or miner in PoW) has full control over which transactions are included in the block. They either solved the proof-of-work challenge or were randomly selected as per the rules of the proof-of-stake mechanism. Therefore, they get to pick the entire contents of the block (which transactions update the database).

The theory was that validators would scan the mempool (the waiting room for pending transactions), and they would pick the transactions looking to pay the highest gas (a congestion fee, or bribe, to include the transaction in the block). Since there is only so much room in one block, the validator should prioritize the ones paying the most.

The rationale for the design was a focus on censorship resistance. The original developers wanted a way to make sure that the network as a whole would be financially incentivized to not exclude anyone from the network. Since anyone could join and take turns being the proposer, even if one of the proposers didn't want to include a certain transaction, the user could always wait until the next block, and they'd get in. The chain was "crypto-economically secure" because it was open and censorship resistant as long as people operated on lines of not burning money (validators are leaving money on the table if they don't include the highest-paying transactions).

Normally this theory works just fine, and it helps to solve censorship concerns for Bitcoin and other non-programmable chains. But there is an issue. This structure of fee markets only works as expected when the transactions are just transfers.

For Ethereum and modern blockchains, the types of transactions taking place break this assumption. Decentralized finance took off, and entire orderbooks, derivative structures, financial tools, and trading platforms were built on-chain. Validators quickly realized that it was no longer optimal to just include any transaction in any order. The specific position of each transaction mattered, and by reordering them or even inserting their own, they could extract much, much more value. When they do this, they capture MEV.

To explain, let's imagine our old friend Jay wants to make a trade. He's feeling a little lucky and has a hunch that the president will try to get the MELANIA coin to go higher. To bet on this, he goes to his favorite decentralized exchange's website and places an order to buy 100 MELANIA coins. It's a market order, so he'll just take them at whatever the current price is. As of the time he clicks 'buy,' it looks like there's lots of liquidity, so it shouldn't be a problem.

Unfortunately for Jay, the validator can see this transaction in the mempool. These validators are holding a spreadsheet and passing around "updates" that they, and users, hope will get included. To take advantage of Jay, the current validator selected to update the database does what's called "a sandwich attack." He places his own transaction to buy MELANIA right before Jay's to push the price up, and then he's the one selling it to Jay at a higher price. Right after the trade, he sells the rest of his tokens, which pushes the price back down. So, if Jay were going to buy at $10, the proposer buys everything for sale at $10 and sells it to him at $11. If the validator does this correctly, he is guaranteed a risk-free profit, and Jay receives a worse price on his trade. After the fact, Jay looks at his screen. The price was $10 before his trade and $10 after his trade, but somehow, he bought it for $11...

Since 2020, over $2 billion in MEV has been extracted on Ethereum alone, with billions more across other blockchains.[i]

whack-a-mole problems

This is not a unique phenomenon. In traditional finance, the concept of front-running has existed as long as finance itself. The broker taking an order on the phone could buy for himself before filling the client's order. The parties running the exchange website could theoretically monitor orders and jump in front of trades. Since asymmetric information of this kind is hard to control for and clearly benefits insiders, the US, like every other nation, has banned the practice.

Just like insider trading, front-running deteriorates the quality of financial markets, and if people don't view markets as fair, they don't participate.

These regulations don't exist in crypto. The difficult part of "insider" definitions is that it's impossible to codify. You need to have it be a subjective definition where parties can determine when something is fair or not. Since code has no court system or jury to deliberate on which actors to punish, it makes it difficult. Every change or market adjustment needs to be handled directly within the code.

For Ethereum specifically, they knew MEV and insider trading ruined their credibility. They needed to make a fix if they were ever going to compete with traditional finance.

They quickly identified that the problem stemmed from validator revenue being dependent on their ability to order transactions. Everyone could see that if one validator was better at reordering the transactions (and inputting their own), they would make a lot more money than someone who just picked the highest-paying transactions in the mempool. Ethereum researchers played out the game theory and correctly saw that the outcome would be that all of the validators would eventually just be run by those who were the best at capturing MEV.

As with most "decentralized" solutions, the way to fix any problem in a neutral way is to just introduce another market.

To explain, blockchains have multiple markets.

The first market that every chain has is a market where parties pay with electricity (or stake) for the right to be the miner/validator for a block. The next is a separate market inside of this role that is choosing the contents of a block. The reward for being a validator became block reward + transaction fees + MEV, where MEV is dependent on each validator's technical know-how in ordering transactions.

To attempt to "fix" MEV, the developers attempted to split the roles of validating (e.g., running the database software) and transaction ordering for profit. To do this, they let validators auction the content ordering portion to the highest bidder. Now, there would be a separate group of "builders" (people who specialize in choosing the contents of the block) who could compete separately from the miner/validator process. They would bid on the rights to order the transaction, and their profits would be driven lower by competition. The auction proceeds would then be split between the validator and the protocol itself, so theoretically it could be used to fund things like public goods and the price of the token. For everyone but Jay, who still had to pay a higher price for his tokens, it sounded great.

builders as scavengers

To give an example of how this MEV fix works, imagine there's a farmer's market. Each day, delivery trucks bring produce from the local farms to sell at your stands. Dozens of old trucks go across your bumpy road throughout the day, and lots of the fruit is scattered across the road.

The trucks in this analogy are the user's financial transactions, the road is the chain, and the fruit is value. Just due to the setup of the chain, value is leaked.

Over time, an economy is created as people stand by bumpy parts in the road to collect fruit. A few guys (professional operators) have the best binoculars and have created complicated methods to view trucks coming and predict where fruit will fall. As a result, they get the vast majority of the fruit.

Now, most people would agree that the scalpers should just give the farmer back his fruit. If that's not an option, we should fix the road, come up with a better transport system, or even just leave it to the birds. The decentralized approach, however, is to make a market around the loose fruit.

The "add-a-market" solution is the local government saying, "We'll just let people bid on the right to stand by the road and collect fruit." The person driving the truck gets part of these auction proceeds, and the rest will go to the local schools.

"Finally," the community says, "the money would go to the trucks and not leak it all to those scavengers." So, at the beginning of every day, the person who pays the most money at the local government auction gets to stand by the road and collect the fruit.

auctions as democracy

Depending on the definition of success, auctioning off MEV on Ethereum worked. Money was funneled back to the protocol.

But there were unintended consequences.

At the start, it was a completely open market for people to come and bid for their opportunity to make money. The bidders were usually individual developers deep in the weeds of financial trading or smart contract development, and they found countless new ways to create MEV opportunities for profit. From sandwich attacks to other race conditions that could net them a profit, this very quickly became a highly sophisticated operation.

As with most markets, though, there was no requirement regarding the distribution of the auction. Soon after introduction, professional traders and firms began to get involved. Since the "best" in terms of making a profit could win the auction every time, the market quickly became very centralized. [ii]

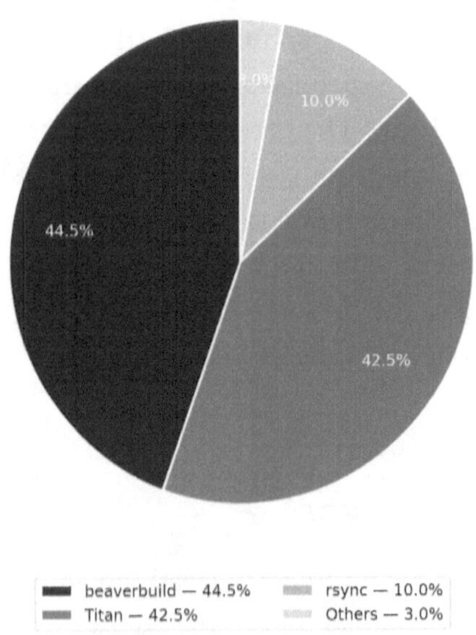

Ethereum Block Builder Market Share, March 2025.
Author's own figure.

IT TURNS out that a handful of firms were much better than everyone else at this. They quickly dominated this winner-take-all market, and the distribution of who gets the rewards is anything but decentralized.

Additionally, the user was still being front-run.

giving up your nodes

Ethereum users (the people like Jay getting front-run) knew that they needed another fix. As a result, the next fix came from the bottom up. In order to protect themselves, users started to use private mempools.

Private mempools are centralized databases run by a series of validators.

Since Ethereum the chain wouldn't build in the judge and jury, users outsourced it to private companies. These companies run validators and agree not to front-run your transaction for a small fee. Now, instead of just broadcasting your transaction to all the other nodes, you simply send it to the trusted validators. Then when it's their turn to update the database, they include your transaction (without front-running it) for a base fee. This might be ok on the surface, but in the long run, it hurts the integrity of the chain.

It again centralizes the chain around private organizations.

As of 2026, over 45% of all transactions skip the public mempool in favor of private ones. Now, instead of even having an open, yet centralizing, skill game as the critical factor, the game has turned into getting access to private transaction flow.[iii]

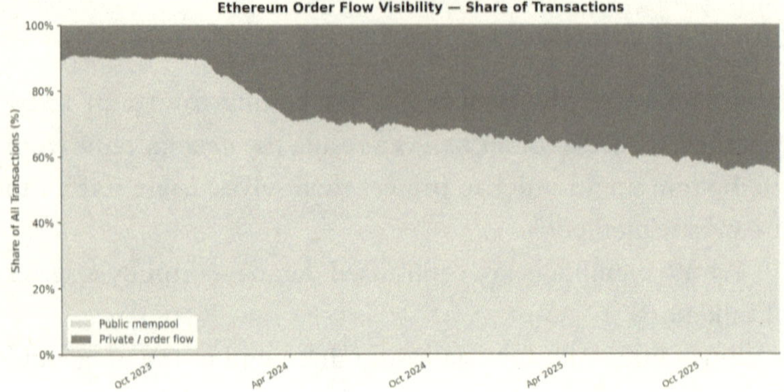

Ethereum Order Flow Visibility: Share of Transactions Routed Through Private Mempools, 2023–2026. Author's own figure.

In our truck example, this would be the farmer's neighbor opening a private toll road where they agree to give you your fruit back if you pay them a fee. Now the system looks more like a racket where you pay the one guy running the road to not steal your fruit.

Users wanted decentralization and censorship resistance, but the market structure optimized for being codified rather than an actual outcome. Since the market created to order transactions favored builders with certain skills, it changed the entire landscape. New markets were set up to try and fix the problem, but each fix introduced different problems. Auctioning off the rights to order transactions led to centralization because it still kept the skill advantage. Privatizing the mempools created benefits for well-connected parties with economies of scale.

Just as with the fuel standards, enacting market regulations to achieve a certain goal can be tricky. If the actual metric isn't "total MEV" or "total carbon" reduced, the market can often miss the entire goal of the regulation. Additionally, the means by which a market achieves the targeted metric are anything but straightforward. If there are values that must be preserved (e.g., decentralization), then they must be specified and controlled for.

To give credit where it's due, the markets did help in the short run. Money was given back to the validators and chain. But it led to unintended consequences.

This isn't an issue of parameter adjustment. This is a feature of markets.

Even if it's not just outright captured and monopolized, the ripple effects of changing these structures are always unclear. It's not that markets shouldn't be used; it's that they need to be constantly monitored and adjusted toward a goal if they're ever going to work. A distribution needs to be enforced, and more importantly, the spirit of the distribution needs to be enforced. There is not and never will be a perfect solution where you can throw away the key and never touch it again. There is no solution to be found in code, only temporary reprieves. It has and always will be gamed. Constantly throwing out attackers and keeping the system fair is hard work, and it always will be.

seeing the future

Ethereum is one of the best chains out there. They know the system is a work in progress, and more importantly, always will be. Whether it will ever be spoken of or admitted is a different matter. Rather than just ossifying and letting the market monopolize or outright fail, Ethereum continues to write up solutions. There are new ideas with new flaws and tradeoffs, but that's ok.

Other blockchains are also trying promising options. *dYdX*, a chain for derivatives trading, opted for just measuring the historical MEV a validator captures, and then the community can vote to slash the validator after the fact if it's seen as extracting value.[iv] This solution introduces a governance procedure (e.g., a vote by other validators and the community) to act as the jury in the case of a dispute.

Even more extreme, other chains have just told validators and sequencers not to do it. They see that the validator set is centralized; everyone knows who the validators are, so just tell them not to. Since they want the chain to succeed, it's in everyone's interest to go along with the idea. They don't pretend to be decentralized to hold onto some belief of compliance or user-enforced ideals.

Yes, there are trust issues, but finding the balance is part of the solution. Which parts of the stack can be treated as traditional companies subject to nation-state laws and which should be open infrastructure are decisions that we're still figuring out.

The good news about crypto is that these communities and companies can move faster than the government. The failure of fuel standard regulations has been known for years, and yet they continue to stand due to the long list of bureaucratic challenges in getting it fixed. Crypto can fix these problems in a much timelier manner. Different chains are trying different things, and the entire space is sharing results. It would be amazing if the technology that sought to remove the government from money actually removes the government from its role in enforcing social consciousness in our financial system.

The industry shouldn't let a belief in neutrality prevent them from making necessary changes.

In the context of fuel regulations, it's obvious that the regulation didn't work out as intended. Of course, the regulations could have been designed to be more robust, but the larger takeaway is that any market will find a way around costly regulations. No matter what the parameters were at the outset, the market and regulations would need to have been adjusted.

For crypto, it's no different. Granted, these systems are tweaking the parameters faster than traditional governments, but they're often still under the assumption that they'll find some perfect setting. They think they can balance all of the variables and then be good to run for eternity.

We're all slowly admitting that markets can adjust human behavior; they just need to be steered if we plan on moving in the right direction.

~

"Bennet! Great to see you!" said a tall brunette.

She ran up to give him a hug.

"You too," replied Bennet as he awkwardly accepted the embrace. She was a familiar hugger in the space, but touching of any sort still made him uncomfortable as he instantly felt that their relationship was inappropriate.

He pushed down the thought.

Her company ran a bridge that Dojima used. They had met a few years back and saw each other a few times a year at these kinds of events. She was also the in for Bennet to all of the good private events.

"How's the conference going? Did you speak yet?" she asked.

"It's good. And not yet," said Bennet rather dryly. Not that she did anything wrong; he just had that glazed-over look that came from networking and nonstop coffee for the past two days.

"Anything spicy?" she said with a smile. "I always like to find out where you see the space going."

"Yeah, you know we're doing a big push to get some RWA players over on our chain. It's a little different now, but just going over how I think it's gonna work out...or not work out probably even more," he laughed.

"Nice, looking for…" She stopped as an older man in a suit walked over. She gave him a hug and began, "Oh Bennet, this is..."

"Mr. DiPolenza?" said Bennet, interrupting, confused to see his former boss here.

He gave a nod and then reached out to shake Bennet's hand.

"Great to see you," he replied, smiling.

"Christopher is an advisor for us and works with Cyborg," chimed in the girl.

Cyborg.money was the far and away leader in the derivatives space that Dojima occupied. Bennet quickly judged them for hiring

TradFi advisors. At the same time, he knew and, from a business perspective, respected what they were doing.

"Glad to see you got in the space more," Bennet said.

"Yeah, it's been a lot of fun connecting these young folks to the industry," Mr. DiPolenza said. "Conferences are a little bit more fun too, if you know what I mean."

There was a pause.

Bennet stared at him. He was still processing his feelings. His territory was under attack. Crypto had clearly failed. Here was a man that literally represented traditional finance, and he was now being paraded around as a trophy by one of the winners in the industry.

There was no way around it. They weren't overthrowing any system. They were becoming the system with the blessing of the old one.

Mr. DiPolenza coughed and brought Bennet's mind back to the conference.

Bennet nodded in case they were waiting for him to respond.

The girl picked up the conversation: "Are you both coming to the afterparty we're putting on tonight? We're throwing it with Cyborg, and you won't want to miss the musical guest."

"Of course, wouldn't miss it for the world," said Christopher.

13

GOVERNANCE OUTCOMES

"What is a country? A country is a piece of land surrounded on all sides by boundaries, usually unnatural. Englishmen are dying for England, Americans are dying for America, Germans are dying for Germany, Russians are dying for Russia. There are now fifty or sixty countries fighting in this war. Surely so many counties can't all be worth dying for.'

-Joseph Heller, Catch-22

It finally happened.

In 2005, Evo Morales became the president of Bolivia. In a nation where over two-thirds of the population was Indigenous, he was their first president.

For the centuries prior, Bolivia's government was controlled by a small group of elites. In the 1500s the nation was colonized by Spain and run by administrators who controlled the territory on behalf of the throne. The Indigenous people, as well as African slaves, were used by the empire to mine silver. Over time, the nation turned into one of two separate classes based on ethnicity.

In 1825, when the country finally gained independence, the

Spanish rule was simply replaced by the new elites. They controlled the land, the natural resources, and even the so-called democratic government.

Like many places in the world, Bolivia was rich in natural resources. At the same time, most of its people lived in poverty. Only five percent of landowners controlled 70% of farmland, and the Indigenous communities that lived on and worked the land for centuries owned almost nothing.[i] Both the mines and profits were foreign-owned, a fact that most Bolivians condemned. In their minds, the arrangement had led to nothing but environmental destruction and extraction.

Morales promised something different.

Inspired by Che Guevara, he wanted the people who lived on the land to have a say in what happened to it. He differed from Che in his preferred method of using an election versus an armed revolution, but he viewed himself similarly as a revolutionary, specifically one that would be taking power back from the oligarchs. At the time, though, getting elected seemed just as difficult as starting a revolution.

Until 1952, Indigenous Bolivians couldn't vote unless they were literate and owned property, requirements that disqualified nearly all of them. Even after voting rights expanded, power remained concentrated among a small political class that rotated the same families through government positions.

Tension had been brewing for decades. Previous movements demanding land reform in the 1990s were met with violence. Despite this, there was still hope that the movement could work through traditional political channels rather than a full revolt. Morales was their best option, and he ran a grassroots campaign that focused on nationalization and Indigenous rights.

After the largest election turnout in the country's history, he won.

Within months of taking office, he took control of Bolivia's natural gas fields, declaring that the resources belonged to the Bolivian people, not to foreign corporations. He passed laws to redistribute land from large estates to Indigenous communities. He even rewrote the constitution to recognize traditional communities and their legal systems. His policies were widely popular among the Indigenous and poor.

But the fight was far from over.

As soon as Morales took office, his plans were stalled. There was widespread pushback throughout both the changing of the constitution and the implementation process regarding taking back control of both land and resources. Like many revolutions, the counterprotest came in the name of freedom, specifically promoting decentralization and local autonomy.

The richest regions were the most opposed. In addition to just being wealthy, the landowners were almost as importantly not Indigenous. They came from a handful of European families who had ruled since their nations took control of Bolivia. Morales's land reforms, namely giving the land and natural resources back to the people, threatened them directly.

As a result, these landowners declared themselves autonomists. They demanded "departmental sovereignty" and the right of each region to control their own resources.[ii] They held meetings on regional autonomy, organized strikes, and even threatened secession. They argued that local communities should control their own affairs, make their own decisions, and resist centralized authority.

Despite an overwhelming victory in the election and a clear referendum by the majority of people in Bolivia, the changes came in the form of compromise found over many years. Foreign control of natural resources was given up, but only at negotiated prices and often with lucrative servicing deals for the corporations. Land was redistributed to the Indigenous people, but it was primarily land that was state-owned and unproductive. Additionally, production quotas were put on the land, which, if not met, would subject the land to auction.

It wasn't the complete socialist revolution that many had hoped for. The breakdown between the goals of the people and their execution was obvious. Even with support from the highest levels of government, change was hard.

Evo's policies and the movement he started were popular for decades to come. He served as president for 14 years, one of the longest-serving leaders in Latin America. Although he didn't accomplish every goal, and compromises were made, the shift towards Indigenous rights and nationalization of industry did happen and has been seen as largely successful. During his tenure in office, the GDP of Bolivia quadrupled.[iii]

A COMMON THREAD throughout all governance and social coordination tools is the breakdown between those who make up the system and those who hold the power over that system.

There's a disconnect between the owner and the employees.

There's a disconnect between the investors and managers.

There's a disconnect between the customer and the corporation.

There's a disconnect between the people and their elected officials.

If there's no underlying social contract that enforces behavior, over enough time, everything turns into a game of who can take advantage of who. If employees have no religious dogma driving them towards hard work, they take breaks, pick up another job, or just slack off unless the manager monitors them. When it comes to executives, they take large portions of the revenue and leave the shareholders with less. When it comes to investors or board members, they extract money through debt at the demise of employee welfare and the company's long-term stability.

This phenomenon is a well-studied concept known as the principal-agent problem. It's when there's a conflict of interest between parties with differing roles and levels of information. The agent (e.g., politician, CEO, etc.) is supposed to act on behalf of the principal (e.g., citizen, employee/investor, etc.), but their own self-interest causes them to act in a manner that is less than ideal.

The classic example is an executive hiring his sons to run the company or paying himself an outrageous sum at the detriment of the company. The investors and employees are dependent upon those in power to execute with often limited means for override. Similarly, if citizens elect a leader with a clear referendum to fix healthcare or fund education and nothing happens, it feels as though the agent is simply dishonest and corrupt, but in reality, the structure of the system itself is broken.

The obvious answer is just to remove the ability of the agent. The process by which this happens is automation and markets.

If every piece is automated and turned over to mutual tradeoffs, there is no conflict of interest. We can all just be contractors and work for ourselves. We can lower taxes and just assume that healthcare and education are better served as private goods.

If everyone just assumes that self-interest is the only way and that taking advantage of the situation is the best outcome for society, then you no longer have a principal and agent. Everyone is both. No one assumes a position of dependence.

This is the logic behind privatization and unregulated finance. Crypto is just an extension to the final degree.

If the government can't properly manage a library due to

fraud and waste, just cut funding for libraries and hope that the free market will deliver an affordable bookstore.

If a company isn't treating its employees properly, just turn them all into contractors who do individual jobs for one-time payments.

The Federal Reserve was seen as not fulfilling its role as a good steward of the money supply, so the community just hard-coded it.

There's no longer an ability for the principal to take advantage of the situation because the role has been disintermediated.

Unfortunately, automation and markets are far from a panacea. Whether it's PoW, play-to-earn, or even transaction ordering, the systems that automate or codify themselves become captured or completely miss their objective. The result of aggregated individual decisions is rarely what anyone would claim is ideal.

Crypto realized this early on. Ossified or unchangeable systems like Bitcoin were limited. They would very quickly become obsolete or captured. In the cases where they did try to implement fixes, the opt-in structure of the upgrades by fork created even more opportunities for extraction by agents or those with insider knowledge.

Blockchain-based systems with no upgrade mechanisms are akin to real-world laws with no means of ever changing. Imagine if instead of upgrading the constitution or our interpretation of the law, citizens just abandoned the legal system and currency every 5 years and restarted from scratch. There's a tempting lure to this system, and there might even be some benefits, but eventually people just want a stable society where they can plan on something being around. The instability and need to constantly keep track of the changes create more damage than they solve.

The broader explanation is that these market-based mechanisms turn every interaction into mutual tradeoffs between two individuals. "X for Y." The feedback and discretion of management, the views of others with X or Y, or even the input by the broader society, are minimized or removed. If X becomes monopolized or all of Y is destroyed, the options are limited. Workers can stand up to a bad employer, but the rights are limited if we're both customers and contractors for some controlling entity. Any effort to enforce a desire is met as an affront to the rule of law.

"You don't have to participate," claims the beneficiary.

In the few systems that do allow changing of the rules, the situation is largely reverted back to the old principal-agent problems. As with Bolivia, sometimes the agent can't control the situation even if they want to. If society wants to redistribute land rights, tax wealth differently, or change how a corporation acts, it often requires years of legal and literally physical threats to make it happen.

So, there seems to be some unfortunate decision that needs to be made between a captured market or a captured agent.

governance structures

Crypto protocols needed a way to use markets but also accept social input and overrides in these systems. As a result, structures were created to alter and update the code to listen to the social demand. This is crypto governance.

Protocol governance, whether it's a decentralized autonomous organization (DAO) or a different arrangement, is functionally just a cryptocurrency wallet that holds funds and operates based on a codified rule set. Like a traditional LLC or corporation, it gives the "organization" a bank account (a crypto address) and rules regarding shared ownership and abilities. Imagine a bank account with programmable rules that determine who can send funds (e.g., only if everyone in the organization signs off on it).

Although the first instance of a DAO was a hedge fund, the main reason governance took off was for changing the rules or updating the code. Developers knew that interoperability was limited if each system might upgrade and deprecate the old ones. If a system wanted to exist and change in some manner not foreseen, it needed a way to accept input from the stakeholders on what to do. And so, they set up methods to listen to the broader social layer.

Unlike Bitcoin or Ethereum, which are similar to databases, cryptocurrencies that have governance structures look more like companies or even nation-states in the way they operate. The only differences are that they tend to be borderless and transparent, but even those qualifications are blurring with time.

The crypto space is evolving and learning as well. They've realized that the larger and more bureaucratic it is, the slower it is to adapt and thus less likely to be able to compete in the hypermarkets of modern finance. The faster they are, though, the more likely they succumb to some mob decision or have some central party take control.

Some projects have opted for faster, more centrally controlled governance systems, but others have opted for slower-moving systems that give more protections to current users. Just as with traditional governments, the issue is the tradeoff between protection of those in the current system and the effectiveness of the change.

To explain through a real-world example, look no further than the governance decisions around wealth taxes. If it takes effect immediately after a vote in Congress, anyone with capital is at risk of having their funds frozen at a moment's notice. This could lead to anyone with wealth moving their assets into something the system can't control, a negative if they're removing investment from the economy. On the other hand, if the vote takes 5 years to fully implement, most of the wealth then will be shuffled to overseas accounts to avoid the tax. Both are potentially dangerous.

In the same way for crypto, if governance can upgrade a system to make it faster or more capable, this is considered good. The problem is that there are incumbents who don't want the system to upgrade. Imagine if a proposal for more inflation in the system comes to bolster research and usability of a system. New users and use cases might be very excited about the initiative, while the current market leaders and token holders may push against the vote.

There's this constant debate that happens over who the system serves. Does the system exist to help those currently in it, or does it exist to help the future users too? Does it exist to serve the current businesses and owners, or does it seek to grow itself?

Even if just as a last resort, enabling the system to alter the distributions of a codified or market outcome without revolution (or fork) is important. Nation-state governments realize that being able to vote through changes to the government, even at the most core level, must be possible. If not, the alternative is violence. History is full of revolutions where captured distributions require a bloody coup to correct themselves. The ability to vote through debt forgiveness or try a new system would have likely been preferable to everyone involved.

The codification and automation of the governance system can give credence to the people of a society.

When the market eventually becomes captured or the distributions undesired, having an outlet to both voice this dissatisfaction and have the collective input actually result in change is powerful. Compared to the compromises needed in the Bolivian referendums, automating redistributions can give power back to the people if the wealth and value are subject to the rules of the system. Just vote in new rules, and the distributions change.

automated trustees

A few years ago, a friend of mine became a trustee of his parents' estate. The man, then in his mid-20s, was something of a failed entrepreneur, and in the eyes of his parents, clearly not too responsible. Rather than just leave him the asset when they passed away, the parents had set up an elaborate scheme where he would get certain distributions based upon life events such as graduating college, getting married, and holding a job for a certain number of years.

The parents were doing their best to ensure their son's success. They just wanted to set up incentives to put their son on the right track. And trusts like these aren't out of the ordinary. As older people live longer and accumulate interest longer, more estates exist, and trusts like these are becoming increasingly common.

It sounds nice, until you play out its logical conclusion. What happens if the majority of wealth is locked in these trusts?

Now, rather than organizing society around what's best for society at that moment, society would be doing the bidding of the automated trusts. Assuming we can't change them as crypto would posit, we've literally set up a society of the living serving the wishes of the dead.

Immutable contracts are choices. The real resources, the work being performed, the houses and property...they all simply exist. If the distribution of society's value isn't one that works, the living should change it.

governance versus the market

The existence of governance in many ways runs counter to the entire purpose of some aspects of crypto. It puts forward the impossible tradeoff space between whether crypto should limit capture (make everything opt-in via immutable systems) or automate acceptance (force through changes for efficiency and transparency).

Rather than just leaving the outcome up to the result of market tradeoffs, governance, when in total control, can change any piece of it. It's both the creator of new markets and the protector of old ones. Through well-implemented governance, parties can voice social consciousness for what really should be the outcome of a social system. Whether it's an admission that a technology needs to be upgraded, an incentive needs to be tweaked, or a distribution needs to be changed, firm organization and governance are some of the most powerful forms of regulation on any market.

The problem clearly is finding the right governance structure. The idea of forcing a project to get feedback from external groups can keep it aligned and prevent market failures from destroying the purpose, but it too can expose the system to a different kind of capture.

The same system that can vote through values-based distributional changes can also be used to vote through nefarious distributional changes. If a system can vote through wealth taxes, it can also vote through theft by an oligarch. As we've seen in crypto, the system that can redistribute inflationary rewards can also direct them unfairly to insider interests.

Many of the structures in crypto don't help much in this regard. DAOs and crypto governance structures are finding it's hard to align incentives in a world with hyper-financialization. If everything is on-chain, even voting mechanisms are up for purchase on hyper-financialized blockchains. Many attacks have already happened where funds are borrowed to vote through malicious proposals. If power is constantly for sale, it's hard to keep stakeholders focused on the long-term value of the system. If parties can sell their tokens and exit, this doesn't lead to even the same level of robustness as traditional corporations.

And this is the balancing act that crypto is attempting to solve. When automation fails, we don't need to remove automation entirely; we need to create methods to add humanity back in such a way that forces interaction and accountability into the system. When the laws are prejudiced, the distributions no longer justified, or the regulations outdated, the system needs to update itself. It's not breaking some natural law to admit that something isn't working. Just because someone enters into a contract doesn't mean it's valid. Just as with the society controlled by dead trustees, if the current participants in the system want to change it, they have every right to.

Figuring out which pieces are subject to the automation of markets and which pieces should be handled by social decisions is a wide-open design space.

Luckily, some structures with promise are emerging.

experimenting with governance structures

One example of an innovative corporate governance structure is that of Hyperliquid.

The company is a crypto derivatives exchange. A popular and crowded space, Hyperliquid gives users ways to make leveraged bets on the price changes in both crypto and other assets (e.g., stocks). They created a unique market structure for trading, but like most crypto projects, the more intriguing aspects of the company have to do with their governance decisions regarding the distribution and usage of their token.

The major structural difference between this token structure and traditional equity is in how it's used going forward. Token issuance and buybacks are both controlled by code. For the issuance, a large bucket (approximately 39%) of all tokens was set aside to distribute as rewards to users. This is a subsidy to users. By giving liquidity providers and traders extra rewards, this tightens spreads and allows for cheaper financial trades.

For buybacks, they too are automated. Rather than accumulate all revenue into an omnibus corporate treasury to be used at management's discretion, a portion of the trading fees (revenue) goes to buy HYPE tokens from the open market. This creates a cheaper product for users (promotes liquidity), but more importantly, it removes the principal-agent problem.

In traditional finance, equity is fixed and only changeable at the discretion of the board. They might do buybacks or issue more stock for a capital raise, but it's relatively limited, and there's a massive conflict of interest. The employees and users rarely see the benefits of the buybacks. The structure is designed to enrich those that control it, namely the board and upper management.

With Hyperliquid, the managerial discretion is taken off the table.

This automation of revenues has the potential to change how all businesses and governments operate. Imagine if revenue could automatically make products cheaper for users. All users who buy a product could be reimbursed their share of the profits for the company over the next month. If the market is healthy, competition could force companies to compete on this lever, literally driving prices down for consumers. A regulator could measure the rate at which the profits are passed back and get an excellent measure of how efficient or competitive a market is.

For traditional governments, imagine if all government documents and files were controlled by code. FOIA requests and documents that are voted on to be unclassified could be automatically executed. This would be a positive for the transparency of the system.

In the case of Bolivia, imagine if the currency had the ability to be programmed. Any revenues from natural resources would always be subject to national control because they would use the currency. Implementation problems become a technical problem of how to automate them with proper governance controls.

social acceptance in the real world

Markets, automation, and social input are all necessary. Hyperliquid automates revenue and token distribution, but they still hold the ability to change where the tokens go. The entire market can be updated if the community chooses.

To push back on crypto's goal of immutability, the automation or "codification" is not the missing piece. It's this balance and our ability to tailor these structures that is the true unlock.

Linking pay to quarterly performance, providing stock options, or even more subjective bonuses have all been used in the past, but they've also been abused and captured. The truth is that more explicit codification of the rules doesn't make this better.

Do we really want to be a society of trustees? Where we enforce automation and where we provide nuance and discretion is a decision to make every day. If we wouldn't opt in to the system now, we shouldn't continue to pretend the distributions make sense or should be honored. Markets get captured, distributions concentrate, and inequities are inevitable. The important piece is the human element added in after the fact.

When you give democracy an automated option, you remove the failures of the agent, and yet you still give the principals a collective voice. Imagine if the people could kick off a vote where, if a two-thirds majority votes in favor, the citizens can override either Congress or the President? There could be an actual democratic check on politicians not doing what they say they are going to do.

Adding in social overrides without revolutions is powerful. Crypto and code can make trust more likely. It can lead to a continued reexamination of whether the current corporate structure is working and can help citizens update the design of their market systems.

We shouldn't treat market outcomes or capture as inevitable. Governance can be the counterweight to automation and financialization if we continuously work to measure and find distributions we all, direct market participant or not, would opt into.

~

Bennet leaned back in his chair as he stared at himself in the empty Zoom meeting. He needed a haircut. He needed to clean his office. He needed to post on Twitter, write a new thought piece for the space, or do a podcast. He also really felt like he needed to motivate the team. It had been forever since he gave a values pep talk.

For another time, he conceded to himself.

Josh joined the call.

"Hey, thanks for hopping on man," said Bennet.

"Of course," he said, nodding.

"Just wanted to talk roadmap and some things we can do to get traction. It seems like everything out there is just dying at the moment. I've put some stuff together and wanted to get your initial feedback. I'm sure you know the runway's looking pretty slim with the token down this much, so we'll need to figure out something to keep it going if our holdings don't bounce back."

The two co-founders were relatively calm.

They had these kinds of talks every few months. Bennet did most of the strategizing and PR while Josh kept the thing running. It had been that way for years.

"What are you thinking?" said Josh. "The chain's stable now, and we have a pretty good roadmap for upgrades."

"Yeah, no, that's great. I think we just might need more of a narrative shift if you know what I mean," replied Bennet. "We've got a great team of developers, and we like the tech, but the users just aren't here. The pitch isn't selling the newcomers, and the TradFi partnerships aren't actually deploying anything, so I don't know how long we can wait for that stuff to take off."

"So, pivot to building our own user? I thought we turned that idea down years ago for regulatory reasons?"

"Yeah, maybe," said Bennet before pausing. "I don't really know either."

Bennet felt like Josh was scrolling Twitter as they sat on the call. For some reason it bothered him. Neither one could make eye contact, but that was nothing new. The one-on-one video call where you stare at another man full-screen was never fun, but Bennet felt annoyed.

Bennet continued. "I'll send it over to you, but what's your thought on moving to proof of governance and hyper-scaling the chain so we can pivot a bit faster?"

"To what?" said Josh.

He wasn't quite as up-to-date on the crypto podcasts or forums as Bennet. Another thing that annoyed Bennet.

"Yeah, we just elect five validators who run the chain, and we can keep emissions and sell pressure low. It seems centralized, but we've seen that PoW and PoS are as well, so might as well just pay people for what they run and get rid of the charade. We'll have some distribution for redundancy, but the security is off-chain in a legal contract, so we don't need so much waste. The "stake" can be put toward liquidity and other things if we want to incentivize that."

"I mean, it doesn't seem too hard," Josh replied, clearly up for filling his time with whatever Bennet wanted him to do.

"Nice, this should make the chain more flexible and give us faster finality, which is apparently a sticker for some of the adoption. I think if we just keep getting rid of the excuses, we should be better against the competition."

Josh knew there would be a big write-up, so this was just a formality. They had runway, and it was just another bear cycle. Dojima had been through a few of these, and this one was no different in his mind.

14

TOKENS AS MONEY

If you're unemployed, it's not because there isn't any work. Just look around: a housing shortage, crime, pollution; we need better schools and parks. Whatever our needs, they all require work, and as long as we have unsatisfied needs, there's work to be done. So, ask yourself, what kind of world has work but no jobs? It's a world where work is not related to satisfying our needs, a world where work is only related to satisfying the profit needs of a business. This country was not built by the huge corporations or government bureaucracies. It was built by people who work. And it is working people who should control the work to be done. Yet as long as employment is tied to somebody else's profits, the work won't get done.

-New American Movement Poster (ca. 1970s)

My first interaction with non-state money wasn't crypto. Living in Baltimore a few years after college, I found that one of the bars around the corner accepted the "BNote," a local community currency.[i] As an economist, my ears perked up, and I instantly asked where I could get some.

The bartender, a longtime supporter of the project, sold them to me on the spot. "$10 for 11 BNotes, and you can use them as dollars," he said. So, I got a discount on my drinks just for using a different piece of paper. It was pretty cool.

I did the same thing at the bar the following week, so it worked out for me, but I also got the chance to talk to some of the people who ran it. Similar to nation-state currencies, the goal was to promote the local economy. If you were using BNotes, the theory was that you would spend your money in the city and that this wouldn't give any business to the corporate oligarchs across the country or overseas. The bar would use it to buy local beer, the brewery would use it to buy supplies, and the supplier would pay employees with it to use back at the bar, a nice little circular system.

I was sold, but unfortunately, I moved away from the city, and sad to say, the currency too faded away. It failed to gain traction beyond a handful of bars and coffee shops, so that whole notion of a circular economy didn't necessarily pan out. Largely idle, the system ran into banking issues a few years later, and the founders just let it die. Like most community currencies over the last few decades, it was little more than a pet project. Getting people to use a new currency turned out to be quite difficult.

promises of crypto

Later that year, I found Bitcoin. Crypto, like the BNote, started off with the simple goal of being used as money. The only difference was that rather than aim to promote local commerce, Bitcoin wanted to save its users from the coming collapse of the dollar. Whether the rationale of a coming hyperinflationary collapse was ever realized is beside the point; the early crypto community wanted it to be money. Countless new coins were all trying to be used as the next digital payment rail that processed the world's transactions.

At least for the hopes and dreams that sustain crypto valuations, it made sense. Given that money constitutes one half of virtually every market exchange, crypto holders still reason that capturing even a fraction of this transactional use case should justify an astronomical valuation.

For years, the crypto community tried to get people to use these assets as a medium of exchange. They would get integrations at stores, websites, bars, and ATMs, but usage always remained low, and the technical details of the integrations left much to be desired.

No one today really uses crypto as money. A few people spend crypto on things, but the majority of schemes simply sell the crypto at the moment of purchase. For any crypto that is used in a real-world transaction, it's almost always a stablecoin that's just a tokenized dollar. It's far from the dream. The idea that our whole economy and every citizen would voluntarily move to a currency outside of the control of the US government seems almost nonsensical now.

But there is a bigger story here. Crypto may not have replaced the dollar, but it's actually playing a larger role in an even bigger shift. Finance is replacing money.

For most of us that haven't touched cash in years, we know that the old days of physical currencies are over. But the migration to digital and programmatic finance is even more drastic. The entire definition of money is quickly changing. What we need money to do today is very different from its role even a decade ago.

a story of social distribution

Long ago, there was a small, walkable community out West, connected only by rail line. Like many small communities back then, they didn't use money, and they didn't even barter; they just shared. With only a few dozen people, it worked. Everyone knew everyone. They went with their gut feeling of who needed what and trusted that it would be reciprocated.

If the baker extended a loaf of bread to the farmer, it was remembered, and when harvest came, the favor was repaid. In the rare case of someone breaking their promises, the neighbors worked together to figure out what was wrong. But generally, everyone talked, so it wasn't difficult to keep track of who was keeping the community afloat and owed a few favors.

Then one day, someone found a piece of gold. It wasn't particularly exciting, but word got out, and before you knew it, the town was overrun with prospectors. Hundreds of people came from the East Coast looking to stay in the town and take part in their economy.

All of a sudden, that system for extending favors didn't really work. They needed a solution to keep track of who to trust and extend favors to.

In order to "legitimize" their personal relationships and interactions, the town decided to create physical tokens that represented these favors. The mayor issued each of the original citizens 10 coins. They agreed that every time someone does you a favor, you give them a coin, and conversely, if you do a favor for them, they give you a coin.

"Now we all know which of the prospectors have done good things, and we owe a favor too." The idea was that each person's balance would represent the sum of their contributions to society minus their utilizations of it. If someone has a lot of coins, then society owes them a lot of favors.

The town had invented money.

As the story illustrates, financialization helps when you can't trust someone. The citizens now don't need to know the prospectors or even have recurring relationships with them to interact. If someone has coins, then you can assume that society owes them a favor. You don't need to know anything about how trustworthy that person may be.

The favors of different kinds, in different circumstances, and to different people are reduced to a standardized measurement. "They clearly did something society deems valuable since they have money."

Our town went from an economic system where societal distributional decisions were subjective to one that's formulaic. Politics and personal feelings could be largely removed from decisions of commerce.

As the town would likely realize, though, financialization goes both ways. It is both the result and driver of a low-trust environment. It can help when trust is minimal, but it can also accelerate the dissolution of social ties if they already exist.

The details matter.

A world where people need to record and get payment for every single favor done is not a desirable one. No parent would ask their kids to pay them for their troubles and no child should expect the relationship to be financial. Very few people would ever want to clock in for time with their spouse. Even the introduction of money to the relationship would be off-putting to most.

When my neighbor comes and asks me to use my lawnmower, I oblige, and he takes care of it and then returns it. If I charge him, however, the entire relationship changes. Now he's less likely to top it off with gasoline. He's also less likely to help with the fence line that's technically mine, but he knocks out most of the time.

Money alters the entire interaction.

It erodes trust between parties that already trust one another. It does this by removing the need for a personal relationship. It reduces a complex decision down to a simple tradeoff. It turns a relationship into a business decision, with clear calculations concerning exit and entry. This might be desirable with strangers, but not with people close to us.

But society moved this way. Gradually and then all at once, we all shifted everything but the closest familial relations to engagements that were monetized. As towns grew bigger, we kind of had to. We had no way to keep our communities secluded, and we didn't really want to. The economic benefits of globalization were too great.

creating systems of trust

As money and financialization moved us towards impersonal interactions, trust and ties to a system were removed. The larger question we face ourselves with now is how to add trust back into these systems. How do we create communities that don't dissolve under the market pressure of their own success?

Crypto can actually help here.

To explain how our new systems are changing money to something more beneficial, we have to first understand that money has traditionally held three functions: store-of-value, medium of exchange, and unit of account. It did everything in one asset. It would hold your net worth, be used for all market transactions, and even be the unit that all goods and services are priced in.

This is actually part of the reason money hasn't taken off in crypto. It's not that crypto is a failure; it's that we no longer have to accept this combination. We can split out the different roles of money.

In the age of digital technologies, where my phone or computer can change the price into any unit I desire, the role of unit-of-account is now meaningless. If you want to price your goods in dollars and I want to buy in yen, there is no friction here. Anyone that has travelled overseas knows they can easily display any price in dollars.

The store of value piece is also being disintermediated, not by crypto but simply by digital markets. Whereas in the past, cash and frictions to investing forced parties to sit in the asset they needed for commerce, this is no longer the case. Again, as any traveler would know, you can use your credit card as if it were in the local currency, oftentimes even with no fee.

In crypto specifically, this ability has been extended to any asset. If you have to pay gas (the network fee to send crypto), you can usually do so in any major cryptocurrency. The wallet software will sell the asset you have, buy the one you need, and use it to pay the fee, all within the same transaction. There is often only a tiny fee for doing this, and as a result, you can use any system, and you never need to hold an asset you don't want to.

For the seller too, if they get one currency from my payment, they can instantly convert it to whatever they want to hold. Most merchants who "accept Bitcoin" do this. When you send Bitcoin for the payment, they instantly sell it for dollars. They don't even touch the "medium of exchange."

Pretty soon, the entire economy won't have people holding dollars. Anyone with a direct deposit will simply roll the payment directly into their portfolio of stocks or bonds. The idea that money will be a store of value is now pointless. Why would you hold dollars when you could just as easily use other options like treasuries or money market funds?

Eventually, it's not infeasible that we will see the entire medium-of-exchange role completely disintermediated by simple 24/7 markets. If the asset is connected to the financial system, it can be used. The benefit of being used as money goes to zero.

So, the question now comes in: What exactly is money going forward?

As crypto has shown, money will be your representative value in a community.

Like the favor token in the Western town, it will likely be used to determine who should or shouldn't be given favors in a society. It will represent your contribution within a community, and it will be less financialized over time.

If there's no longer a need for money in transactions, nation-states and crypto will have to compete within the realm of creating something people want to hold. These systems will use their ability to tax or even force the holding of an asset on people looking to interact with that community. This is where trust is added back into the equation.

If parties are forced to hold or store value in a given asset, their economic well-being is tied to governance around that system. For dollars, holding the currency means you trust the issuer and the society around it. If the goal of the asset issuer is providing housing, healthcare, and retirement to its citizens, you better be sure that the citizens feel taken care of and are educated enough to know when to inflate or devalue the money supply. It's nothing more than a tool for coordinating who should do what for each other, and if the broader society doesn't want to honor your token's value, they don't have to.

If a large number of favor tokens are held overseas and then confiscated in some manner your government deems reprehensible, there's no need for that society to continue to accept them. If the majority of the favors are owned by one party, locked in trusts controlled by the dead, or simply distributed in a way that no longer is best for the society that uses them, the rules can change.

The rules, the value, and the trust are social. They form interactions and incentives and can be designed however people want them to be.

If currency issuers and society begin to see these assets as tools to enforce trust, they will likely deploy new measures to get people to hold their assets. If you think of money as a favor token we're giving to someone, more methods are needed to ensure that the parties are still the ones we want our system to interact with. Allowing people to exit with value or creating frictionless transfer of social favors across communities can have tangible negative effects if the ending distributions run counter to societal goals.

Luckily, crypto is figuring out solutions. These systems are decades ahead, because they already exist in a world of seamless movements of financial value.

In crypto, we often talk about creating "trustless" systems. In reality, this couldn't be further from the truth. We're actually trying to design systems that force trust. By using cryptography and economic collateral, developers try to align incentives so you can trust the person and community who you're interacting with.

In the same way our town created money so they could trust their interactions with the prospectors, crypto is the way it is because we have these participants who we don't know, and ideally, don't even want to know. We're trying to force these new parties to fit into our community and do beneficial things for our town. That's the goal: align incentives so they can interact with us and we can open our borders.

The best example of this is staking.

Everyone knows about proof of stake, but the concept is being deployed for things beyond just securing a database. The basic idea is that anyone can lock up some assets (e.g., ETH, USD, TRB, etc.) and then they will lose them if they break the rules.

This is easy to see in the case of database updates; if a validator tries to steal money, the rest of the community will take their locked assets (or it will happen automatically in most cases). But it can also work for more subjective pieces. Many governance systems, for instance, require parties to stake assets in order to vote. Now if someone tries to vote through a bad proposal, they have skin in the game and will lose that value.

Imagine if you had to lock ten percent of your wealth in a zero percent bond in order to vote. If exit with value was not an option, we might make different decisions with our votes. Systems could even weight them by time locked so the longer the stake has been around, the more votes that person will have. Systems could also have very long times to unlock (meaning each voter is stuck with the result for a long time).

Since the asset is staked, parties are tied to the community. In the same way holding a stock requires trust in the company, having people stake monetary assets is an excellent way to get them to show that they are invested in the community's long-term well-being.

For equities, we're likely to see experiments take place with corporate governance in the same manner. Like DAOs, corporations could require shareholders to lock their stock for five years in order to vote on new proposals. Now there would be a monetary incentive to being fiscally responsible and more long-term focused with their vote. Clearly other problems might arise, but it's a better option than allowing short-term owners to exit their position if things start getting tough. Localities might even require corporations to stake assets in order to hire or sell to their citizens. As startups and investors seem to flip companies or ideas for extractive purposes, longer-term staking and ties to an idea and a community will likely be taken up as options in slowing down the negative effects of any technological innovations or market introductions.

With enough of a network, tools being pioneered by crypto governance can be a better way to interact than just financializing a market and leaving the outcome to the invisible hand.

If society changes, the value given by the older society should change as well. Just like with our favors currency, if the entire town of newcomers is forced to serve the original tenants forever, they might be better off just starting their own town (or currency).

It's really an open problem that crypto is also working at solving: how do we update distributions without a violent revolution or existential technological shock? Can we actually create an opt-in system that allows us to escape from the captured institutions of the past? Maybe the key is by targeting value. If it's social, and money leads to power, maybe by voicing our social consciousness in what we hold, we can push some of our values back into the market and even the system as a whole.

designing fairer systems

It's fitting that crypto is still working on building money. The ideas and goals have just changed. Crypto is blurring the relationship between an investment and currency, and the financial world will change drastically because of it.

At a nation-state level, this means that interest rate policies and currency controls will become very difficult to enforce. If you let people invest and move value freely across borders or across assets, the ability to tax and regulate their usage becomes very difficult.

At a company level, it means that equities and currencies will be on the same footing. People will choose to invest and hold what they want to succeed.

Creative tools such as automated issuance and distribution will be key in separating which investments people choose to hold. Where the money goes and whether it benefits the past or future holders will continue to be the main point of contention.

Every time an asset is issued or a choice to hold one asset over another is made, redistribution of value occurs. You're moving value to those that already hold that asset. You marginally push up the price and redistribute value and legitimacy to the asset. Often the term "redistribution" can have a negative connotation, but in most instances, it enables something beneficial. This is the superpower of fiat currencies and crypto: the ability to direct which pieces of society are valuable.

Forcing people to hold or stake a certain asset in order to be part of the community can align interests. This isn't some novel concept. If a startup's employees are paid in dollars, they can just leave and take their money with them. If, however, they are paid with vested equity, the "currency" used in the transaction helps to align interests. Crypto is just expanding the flexibility of this practice.

shifting real resources

Money is no longer needed to be a store of value or even a unit of account. Everything going forward is just an investment and a token for participation. The industry is leading the way in creating pseudo-equity/currency instruments that can be used to incentivize certain behaviors. We just need to look at the results.

Is it causing people to put resources towards productive and wanted outcomes? Or is it exacerbating inequality and removing trust within communities? Is it letting people put their value where their values lie? Or is it letting the wealthy escape taxes and social regulations?

The goal of both finance and money is to shift resources in the real economy.

Digital assets are just coordination tools. Their purpose is to change how people interact with one another toward some goal. The BNote was issued to drive a local economy. It worked if more people spent money within the city.

In the same vein, if there are men with no work, resources that could be put to better use, a community that needs education, or workers that need protection from market disruptions, money can be restructured, and different actions will take place as a result. The issuer of an asset with value can solve collective funding problems. Building roads and funding services isn't a "venture" or investment problem; it's a problem of aligning interests between those with resources and the larger society that gives them value.

~

After a few minutes of modern elevator music, the host finally unmuted and kicked off the X space.

"The man, the myth, the legend, Bennet Cooper. Thanks for doing the space with me," the host said sarcastically.

Bennet had met him a few times before. He'd been around for at least the last two cycles and ran FintechWarriors, a very popular Twitter account and media business in the space.

"Of course, excited to be here. What are we talking about today?" Bennet replied.

"Well, you know it's never a quiet moment at the forefront of crypto Twitter, so there's obviously lots of things we can talk about, but given you are the CEO of Dojima, what's going on in the defi space right now? There was the recent bill that just passed the House that seeks to regulate you guys; any thoughts on it?"

Bennet responded with his scripted answer. He felt like a machine with these kinds of interviews.

"I think the bill is much more stablecoin-focused than anything generic tech like we're doing, but it's super positive for the space. We've been looking for regulatory clarity for years, and it's great that we're finally getting some. I think this is the big talking point in crypto and all of finance right now. Figuring out how the administration is going to act and whether we can compete on our own merits with the TradFi world is the next challenge."

"I think that's the sentiment we've been hearing a lot lately. How does Dojima take advantage of this? And would you say we're in a bear market now?" asked the host.

"Yeah, I mean we're definitely in a bear market. BTC's off its peak by what, fifty percent? The alts like us are down more obviously, which is normal. But yeah, we've been through quite a few market cycles now, and I have no doubt we'll come out stronger from this one too. The market's changing, and we just have to adjust like companies do in any market."

Bennet noted only 22 people in the space. He started wondering if this was a waste of time. Since there was no video, he opened a new tab to look at prices.

The host continued, "How would you say crypto's different, though? Other markets don't have these wild swings like crypto, right?"

"Yeah, I mean, crypto's just more extreme. I think you saw it with the early web too. Early on, you had the cypherpunks of the world who wanted to build one kind of internet, and then things changed with the dot-com bubble and the adoption of cloud and social media. We could finally make apps at the scale that people actually wanted. I think that's just what's happening here. We're moving past the early stages and having some growing pains as we figure out the technology and what pieces truly have product-market fit."

"And what are those pieces?"

"I think it's clearly finance." Bennet continued. "We have almost a decade of crypto disrupting various pieces of the financial stack, and I think the liquidity in crypto is something that's stuck and is clearly better."

"Ha, so you're not sold on the NFT craze coming back. No more digital cats?"

The host was kind of annoying Bennet. He wanted to talk about something interesting. He figured it would be inappropriate to start talking about music.

"I know better than to count things out. But people like gambling, so maybe something even crazier will come back. But overall, I don't think those original cats are going to be worth much; the new guys will push something new and take the mindshare."

"Yes, definitely couldn't agree more," the host said.

The space was quiet for a few seconds.

They were now down to 15 participants.

15

INTENDED CONSEQUENCES

"The purpose of a system is what it does. This is a basic dictum. It stands for bald fact, which makes a better starting point in seeking understanding than the familiar attributions of good intention, prejudices about expectations, moral judgment, or sheer ignorance of circumstances."

-Stafford Beer

I love cooking.

For some reason, I especially get a kick out of making a good dessert. Not because I'm necessarily a sweet tooth or even enjoy baking more than grilling, but I think it's just the infrequency of my attempts that makes it special. It's also so methodical. It's straight measurements and instructions. And the really glorious part is that once you get the basics down and find some good instructions, you get that fantastic joy of it working every time. Recipes are the smell-good version of Stack Exchange answers that you cut and paste, and it magically creates working code.

A few years ago, wanting to push my cooking to the next level, I decided to challenge myself by finding a little competition. For me, my mother-in-law is sort of the "good cook" around; you know, the "world-famous cookies" and "best homemade pie" kind of Polish grandma you'd expect kids in some movie to have. So, I started off hot, right after the prize winner... Let's see who can make a better cheesecake (one of her signatures).

Now I knew hers was good, but to be honest, it was kind of homely. Probably on par with a Cheesecake Factory / good store-bought type of vibe. She'd had the benefit of living in a bubble, and that's sort of the point.

I went online and found an old-world ricotta one that fit the bill: 5 stars, a website that looks like a hobbyist's and not some advertisement, and a decent number of reviews and tips.[i] I practiced the week before the event, and I knew as soon as I tasted the batter that I had myself a winner.

So, fast-forward a bit; on the night of the contest, the family declared it was a tie. Hers was the "classic" taste, and mine was just "amazing, but not a cream-cheese base, which was expected" (how could you ever compare?). The seconds taken were all mine, though, so internally this was a fine victory for my baking skills.

The contest was a success, and I started doing this every holiday against different relatives' hand-me-down recipes and "traditional" favorites. Coming to even my own surprise, every holiday I would get the same results or better.

The internet claimed victory over the hand-me-down recipe.

For so long, each family or community had its own little bubble. Recipes were shared via word of mouth, and oftentimes it was pretty uncontested (how many cheesecake recipes were floating around small Pennsylvania towns in the '60s). These little word-of-mouth bubbles actually preserved the old-fashioned recipe. It was able to survive because there weren't other options.

But now we're online. We're not stuck with grandma's recipe. I don't think anyone these days would even think of calling a parent for a recipe.

Assuming you don't just ask your AI, you go online and look at rated recipes. Comparisons and reviews that have touched the tastebuds of thousands of home cooks each year, with tips and a comment-section-style open-source development of the "best" recipe.

It's hard to beat. The market for recipes has delivered the best out there for the most people.

And this is great. Rather than recipes coming from the "centralized" authority being the person in your social circle with the instructions, the openness to all, say the "decentralization" of the recipes, leads to everyone getting to enjoy a little bit better of a cheesecake. For those economists out there...more utility and less inefficiency, right?

But digging deeper, we can all see where we end up: converged at a global level. A flattening of culture that other authors have written about through decades of globalization. We have a handful of battle-tested recipes that stand up to the test of time and work as the "best."

Of course, you could say that you have more "decentralization" because you can pick from any recipe; people are free to choose, and as a whole, utilitarians would argue, society is better off. But things don't exist within a given moment of utility measured. We have to think about how the flavors of the past, the hand-me-downs, are lost. The "centralized" aspect of one recipe in each bubble was actually better for the "decentralization" of the system as a whole.

And this isn't just recipes. You can see the internet as the "decentralizing of information." Rather than have just one local paper, a meh beer from the factory down the road, your local news, or even one library, you now get it all! No more relying on that pesky post office to send letters, the government-sponsored library for research, or even your grandma for a recipe. Every piece of information, every product, and nearly every experience on the globe is now at your fingertips, and it's amazing.

But have you ever wondered why every major city has amazing restaurants that for some reason all taste the same as restaurants in other cities? Global supply chains and information sharing have led to a homogeneity across cities, and more recently, nations. This has been commented on with restaurants, style, movies, music, and even things as subtle as brand design and naming.

It's not a direct market failure but rather the result of aggregated individual decisions resulting from increasingly efficient systems.

What happens when the new equilibrium for a market is a monopoly (or close to it)? Rather than hundreds of local monopolies, you get a global duopoly. Is this progress? What happens to all those people in the future who might like a different cheesecake? What happens to the experience you get travelling to cultures that differ from yours? And maybe those second and third-best recipes and restaurants should go away. Maybe we should just have the best...It's opt-in, right?

It is on paper, but we all know it's not. When looking at the internet, it's obvious it's been captured, rather ironically, by our freedom to all pick the same thing. Network effects, global communication, and our race for simplicity and a recognized user experience have driven us to centralization.

The simple act of opening something up, globalizing it, making it immutable, or taking the local "state" control out of it doesn't mean you get what you want. Rather than open the whole world up to new ideas, new opinions, or systems, we all decided that it was an excellent time to standardize the structure. Rather than make it easier to travel to family restaurants with different dishes, we decided to export the best dishes and homogenize restaurant experiences around the globe.

It's what the market wanted apparently.

In terms of what we're talking about here for decentralization and digital systems, these "global systems," the internet, the AI, the all-encompassing cloud or database, and the global news network; each one further removes locality and nuance. It reduces every interaction down to a quantified financial tradeoff.

We can't possibly consume it all, so the winning team simplifies it. It gives a "right answer" that's easy to find. Whereas Google led me to 5 recipes, the search in the next-generation AI model doesn't give you hundreds of results with personal reviews; it gives you one. A winner-take-all system that drives us toward a "truth."

Markets are global. Financialization removes the friction of the physical world. Neutrality means we give up any ability to stop it.

It's a trap of our own innovation.

The questions that we need to ask are how we can actually build systems that are good in the long run. How can we differentiate between the initial goals of a system and then the greater purpose being the consequences of what it does? There's a constant fight between short-term business success and the success of our long-term social structures.

Do you really want your new cryptocurrency to surpass the US dollar or replace banks?

Maybe it's fine if our new currency is just for a specific community. I know valuations won't appreciate that take, and it's most likely why we don't see it, but very few of the cryptocurrencies around today have a positive, articulated vision for a future in which they succeed.

For any currency, be it the US dollar, Bitcoin, or any other crypto, total monopolization as the world's one currency is in the long run as undesirable as the globalist-run CBDC the conspiracy theorists rally against. We've all been told to fear the "one world government" as it pushes its tentacles to control us from afar. In fact, Bitcoin and many other cryptocurrencies were created to push back against a currency which we didn't feel represented our best interests.

But what if our interests include both succeeding? The fact that Bitcoin is a minority system is in and of itself part of the answer. Currencies have enormous network effects: brand recognition, perceived stability, and acceptance. Market systems like these monopolize in the long run, and it's the job of a vigilant population to combat this force by creating new options. If we can throw the old distributions out the window by adopting a new one, maybe we can put a check on the oligarchs that have captured the system.

Like the cheesecake recipe, a single crypto winning is worse in many ways than the centralized alternative. The system would be captured, but the narrative of immutability would fight the idea of any change. Don't like the distribution of your new currency? Do you disagree with the wars waged via the inflation in a system? Is the entire system censoring transactions from North Korea and Iran? The decentralization and lack of clear paths for changing the system may make it even less clear than now how to enact change at a national (or global) level. Say what you will about the US government, but at least there's some path to changing the constitution.

Decentralization is hard to define. There are different levels of society that systems interact with. You can have decentralized, democratic control of a monopoly, or you can have private ownership of "competing" firms in a free market.

Pushing things towards markets isn't always the answer if the end state of the competition isn't clear. Are the five validators we choose in an election really more centralized than the two chosen by the open market? If a king is turned into a "democratically elected" president, has the illusion of choice affected thinking around class justice or revolutionary actions?

On a more micro level, we see markets introduced as efficiency-generating machines, solving all budgetary constraints. Government services privatized to save money by allowing the free market to work; an excellent way to "decentralize," right? But then in a few decades, Americans are scratching their heads over how they ended up with private prisons that give kickbacks to judges, national parks that charge access fees, and private military defense contractors engaged in shadow wars.

Decentralization that leads to fewer options and less liability is the worst of both worlds.

Maybe the real decentralized future is fragmented. Maybe one world currency, one world data center, one world social media, and one world recipe are things that should be thought about differently before we just see how fast we can get there. Code is great as an alternative path forward for coordinating with one another, but we need to be careful that we don't imagine it to be the only path. Decentralization and technology should enhance the social layer, not remove it. It should allow democracy to work better. It should allow good ideas to spread and information to flow freely. It should promote transparency and individual freedoms in the broader context, not just in one small (often financial) context.

concluding optimism

Some might look at the examples discussed in this book as a massive success. Early America using land runs helped to distribute land more equitably. Stock certificates of the VOC were a new financial technology that enabled crowdfunding of corporate ventures. Bitcoin achieved institutional adoption and is worth trillions of dollars. The fuel regulations passed by Obama were a massive success in pushing through wanted environmental regulations. The prediction market that correctly informed us of Taylor and Travis's engagement was a revolutionary use of finance.

Others might see it differently.

Land runs were theft from natives. The VOC helped to finance a genocide of an entire island of Bandanese and perform countless other atrocities. Bitcoin failed to provide an alternative currency or shift the power structure away from the US government. The actual implementation of the fuel regulations led to an increase in CO_2 and larger, more dangerous roads for Americans. Prediction markets exploit gamblers in order to benefit insiders and produce societally useless information for social media content.

Technologies that help us collaborate can be good, but technologies that erode existing social structures can be detrimental.

Whether crypto can be looked at as a new technology enriching non-traditional investors and providing cutting-edge cryptography research, or whether it's a series of market failures fueled by gambling addictions, is up for interpretation. If we're going to find solutions for making better systems, though, we have to get imaginative, and we have to admit that failures did and will happen. Whether you want to say it was because of market dynamics in general or because the specific market wasn't coded properly is up to you, but don't allow anyone to tell you that the current landscape was the goal or the intention. We were sold something very different when we bought into the story of crypto, and we owe it to ourselves and our future to be honest.

Despite most of the space running headfirst towards neutrality and financialization, some of crypto shows promise of fighting back. Experiments in governance, money, and market-based regulations aim to increase and personalize social coordination. We just need to be specific about what parts we want it to solve and which parts we need to save for human discretion.

Building "censorship-resistant" or "credibly neutral" applications is not a purpose.

If you maintain a tunnel and it's used to help free slaves or as part of a resistance movement, you should keep doing so. If the tunnel is used to traffic children and weapons for cartels, you should probably stop maintaining the tunnel.

"Anyone can use it for anything" is simply an admission that your platform is open for sale to the highest bidder. It's a relationship with no commitment even to itself.

If what we're building is a tool for the oppressed to opt out of structural institutions that restrict them, then it can succeed. However, if it's a tool that the powerful use to escape the will of the people, this is dangerous. What it's used for matters, not the stated neutrality in the whitepaper.

we'll get there

I still compete in cooking competitions, but I'm trying to just ask around more for recipes. Changing little pieces and moving forward each flavor with some help from the online communities. The new dishes often lead to better conversations about their origin, and they're definitely more fun to experiment with. Maybe finding the efficient path to winning was never the point.

Running from society is not an option, and there's no market design that can solve all of our problems. Code can only have the value which we give to it, and it's time to admit we have the choice.

If crypto will succeed in protecting privacy, sound money, and freedom of speech, then we'll need to convince a large number of people to adopt it and make sure it's accepted and legally protected. But conversely, if you can convince a large number of people that privacy, sound money, and freedom of speech are important, then you don't need crypto.

~

Bennet read the proposed headline in cold disbelief, "*DALIOX Acquires Dojima Labs and Related Intellectual Property.*"

Goddamn, he thought.

This was his baby.

He'd seen this thing grow up for the past seven years of his life, and now he'd be selling it for parts. How on earth was he supposed to give a quote?

Maybe it didn't matter. Maybe his assumptions were wrong, and this was what growing up looked like. Selling was a noble exit after all.

Looking back, he knew he made lots of mistakes; any first-time founder does. He even felt he navigated it all pretty well too. The pivots, the changes, the staying relevant. Few other projects stayed around or had the success they did.

Granted, he didn't get to cash out quite the same as other founders. He made plenty more than at his previous job, but the dreams of retirement were still that. He'd be working for his new boss, a corporate payroll processor, to help them design their new stablecoin.

He figured there could be worse gigs.

Bennet typed: '"We're super excited for this move," said Bennet Cooper, one of the founders of Dojima Labs. "We built some amazing technology, and I'm so proud of our team and community for making it possible."'

He liked that in a nauseating sort of way.

He continued: '"It's exciting to be working with such a respected player in the institutional world and to finally be able to bring about our vision for the future of finance."'

Slightly more nauseating.

He knew the token would tank. He still had some left, but they were down so far already he counted them as worthless.

It seemed no one was paying attention anyway.

He figured he'd need to find another idea that might actually disrupt something. Maybe he could do another startup.

He liked that idea.

He hit send and opened a new document.

REFERENCES

1. Defining Markets

i. Meyers Photo Shop. (ca. 1889). *Land run* [Photograph]. Oklahoma Historical Society, Gateway to Oklahoma History. https://gateway.okhistory.org

ii. Homestead Acts. (n.d.). In *Wikipedia*. Retrieved December 5, 2025, from https://en.wikipedia.org/wiki/Homestead_Acts

iii. Forrester, P. (2024, June 3). Henry George: An exploration of some consequences to taxing only land. *Econlib*. https://www.econlib.org/library/columns/y2024/forresterpovertysolution.html

iv. Land Rush of 1889. (n.d.). In *Wikipedia*. Retrieved December 5, 2025, from https://en.wikipedia.org/wiki/Land_Rush_of_1889

v. Bloch, M. (1961). *Feudal society* (L. A. Manyon, Trans.). University of Chicago Press. (Original work published 1939)

2. Starting the Chain

i. HashrateIndex. (2025, January). Bitcoin mining pools distribution. Retrieved from https://hashrateindex.com/hashrate/pools

ii. Hashrate Index. (2024, December). Top 10 Bitcoin mining countries of 2025. Retrieved from https://hashrateindex.com/blog/top-10-bitcoin-mining-countries-of-2025/

iii. Lightning Network. (2025, December 10). In *Wikipedia*. https://en.wikipedia.org/wiki/Lightning_Network

3. Credibly Neutral Distributions

i. Proof of stake. (n.d.). In *Wikipedia*. Retrieved December 5, 2025, from https://en.wikipedia.org/wiki/Proof_of_stake

ii. Proof-of-stake (PoS). (n.d.). *Ethereum.org*. Retrieved December 5, 2025, from https://ethereum.org/en/developers/docs/consensus-mechanisms/pos/

5. Scaling the Chain

i. Braess's paradox. (n.d.). In *Wikipedia*. Retrieved December 5, 2025, from https://en.wikipedia.org/wiki/Braess's_paradox

ii. What is the blockchain trilemma? (n.d.). *Coinbase Learn*. https://www.coinbase.com/learn/crypto-glossary/what-is-the-blockchain-trilemma

7. Tokens – Commoditized Financialization

i. Dutch East India Company. (1606). *Share certificate, Chamber of Hoorn* [Document]. Westfries Museum, Hoorn, Netherlands. Public domain.

ii. Ricklefs, M. C. (2008). *A history of modern Indonesia since c. 1200* (4th ed.). Stanford University Press.

iii. Petram, L. (2014). *The world's first stock exchange*. Columbia University Press.

iv. Gaastra, F. S. (2007). The Dutch East India Company: Expansion and decline. *Walburg Pers*.

v. CoreLedger. (2019, June 6). What is tokenization? Everything you should know. *Medium*. https://medium.com/coreledger/what-is-tokenization-everything-you-should-know-1b2403a50f0e

vi. Colored Coins. (n.d.). In *Wikipedia*. Retrieved December 5, 2025, from https://en.wikipedia.org/wiki/Colored_Coins

vii. de Soto, H. (2000). *The mystery of capital: Why capitalism triumphs in the West and fails everywhere else*. Basic Books.

viii. What is Peanut the Squirrel (PNUT) and how does it work? (2025, May 9). *101 Blockchains*. https://101blockchains.com/peanut-the-squirrel-pnut/

ix. Altcoin Buzz. (2024, November 20). Dune Analytics report: Most memecoin traders lose money. https://www.altcoinbuzz.io/cryptocurrency-news/dune-analytics-report-most-memecoin-traders-lose-money/

x. LiquiFi. (n.d.). Token vesting and allocations industry benchmarks. https://www.liquifi.finance/post/token-vesting-and-allocation-benchmark

8. Portable Value – Gaming, NFTs, and Open Reputation

i. Graves, S., & Hayward, A. (2022, March 30). What is Axie Infinity? The play-to-earn NFT game taking crypto by storm. *Decrypt*. https://decrypt.co/resources/what-is-axie-infinity-the-play-to-earn-nft-game-taking-crypto-by-storm

ii. Screenshot of Axie Infinity (Sky Mavis, 2018) reproduced under fair use doctrine for purposes of critical commentary and analysis. All rights reserved by Sky Mavis Pte. Ltd.

iii. Priori Data. (2022). *Axie Infinity monthly player count statistics by country*. Retrieved from https://prioridata.com

iv. Data sources: CoinGecko; Kraken Historical Data; CoinLore; CoinMarketCap (2026). Prices are approximate monthly closing values.

v. Rochet, J. C., & Tirole, J. (2006). Two-sided markets: A progress report. *The RAND Journal of Economics*, 37(3), 645-667.

9. Liquidity and Exit

i. CoinGecko Research. (2025). *Crypto liquidity on CEXes 2025*. CoinGecko. https://www.coingecko.com/research/publications/crypto-liquidity-report-2025

10. Prediction Markets

i. @themandalore9. (2025, August 26). *Information markets really are the future. By leveraging gambling addicts, we created liquidity for insiders to trade against* [Tweet]. X. https://x.com/themandalore9/status/1960417142786142392

ii. BeInCrypto. (2025, August 27). Taylor Swift engagement: Did her guitarist game Polymarket odds? *Mitrade*. https://www.mitrade.com/insights/news/live-news/article-3-1071393-20250827

iii. Polymarket. (2025). *Taylor Swift and Travis Kelce engaged in 2025?* [Prediction market]. Polymarket. https://polymarket.com

iv. AsymmetricInformation. (2014, February 19). *[WHITEPAPER] Decentralized Bitcoin prediction markets* [Online forum post]. BitcoinTalk. https://Bitcointalk.org/index.php?topic=475054.0

v. Wikipedia. (n.d.). *Augur (software)*. Wikipedia. Retrieved April 28, 2026, from https://en.wikipedia.org/wiki/Augur_(software)

vi. Futarchy. (n.d.). In *Wikipedia*. Retrieved December 5, 2025, from https://en.wikipedia.org/wiki/Futarchy

vii. CryptoKoryo. (n.d.). Prediction markets [Data dashboard]. Dune Analytics. Retrieved January 3, 2025, from https://dune.com/cryptokoryo/prediction-markets

viii. *Betting markets versus polls: Trump vs. Harris, 2024* [Figure]. Author's original figure. Data sources: FiveThirtyEight poll aggregates; Polymarket prediction market probabilities (polymarket.com).

ix. Byrne, S., & Volpicelli, G. (2026, January 12). An anonymous Polymarket trader made $400,000 betting on Maduro's downfall—and now Washington wants answers. *Fortune*. https://fortune.com/2026/01/12/polymarket-kalshi-insider-trading-prediction-markets-cftc-torres-titus-venezuela/

x. Polymarket. (2026). *Maduro out by...?* [Prediction market]. Polymarket. https://polymarket.com

xi. MarketsWiki. (n.d.). *Banging the close*. MarketsWiki. Retrieved April 28, 2026, from https://www.marketswiki.com/wiki/Banging_the_Close

11. Freedom and Markets

i. Britannica. (n.d.). *Industrial Revolution timeline*. Encyclopædia Britannica. Retrieved April 28, 2026, from https://www.britannica.com/summary/Industrial-Revolution-Timeline

ii. Unknown Artist. (n.d.). *Luddites smashing textile machines* [Illustration, engraving with modern watercolor]. World History Encyclopedia. Retrieved April 28, 2026, from https://www.worldhistory.org/image/17132/luddites-smashing-textile-machines/

iii. Creative destruction. (n.d.). In *Wikipedia*. Retrieved December 5, 2025, from https://en.wikipedia.org/wiki/Creative_destruction

12. Fee Markets - MEV and Skill Games

i. Chi, Y., Wang, X., & Liu, Y. (2024). Maximal extractable value: Current understanding, categorization, and open research questions. *Electronic Markets*. https://doi.org/10.1007/s12525-024-00727-x

ii. Gate Research. (2025). Monopoly in Ethereum block builders and chain abstraction: Profit incentives and innovation in the blockchain ecosystem. Gate.com Learn. https://www.gate.com/learn/articles/monopoly-in-ethereum-block-builders-and-chain-abstraction-unveiling-profit-incentives-and-innovation-opportunities-in-the-blockchain-ecosystem/7690

iii. @dataalways. *Private order flow monitor* [Data visualization dashboard]. Dune Analytics. https://dune.com/dataalways/private-order-flow

iv. dYdX Trading. (2023). *An update on MEV: Catching a bad validator*. dYdX. https://www.dydx.xyz/blog/update-on-mev

13. Governance Outcomes

i. Webber, J. R. (2011). *From rebellion to reform in Bolivia: Class struggle, indigenous liberation, and the politics of Evo Morales*. Haymarket Books.

ii. Eaton, K. (2007). Backlash in Bolivia: Regional autonomy as a reaction against indigenous mobilization. *Politics & Society*, *35*(1), 71–102. https://doi.org/10.1177/0032329206297145

iii. World Bank. (n.d.). *GDP (current US$) - Bolivia* [Data set]. World Bank Open Data. Retrieved January 28, 2026, from https://data.worldbank.org/indicator/NY.GDP.MKTP.CD?locations=BO

14. Tokens as Money

i. Where to get BNotes. (n.d.). *The Baltimore BNote*. https://baltimoregreencurrency.org/cambios

15. Intended Consequences

i. Shiran. (2024, February 29). *Ricotta cheesecake*. Pretty. Simple. Sweet. https://prettysimplesweet.com/ricotta-cheesecake/

ACKNOWLEDGMENTS

I'm grateful to countless people who have helped make this book a reality. This work and my entire journey into crypto wouldn't have been possible without the support of my wife, who's always believed I could figure it out. The book was a journey in itself and over the past few years, I've learned how little I've even scratched the surface on these concepts. Gratitude to all my editors and early readers including my mom, my brother Tyler, my dad, the McPhail family, Nate, Jay Rush, Andrew Ness, Nick Blasco, Jake Matthews, and Reese Flurie.

www.ingramcontent.com/pod-product-compliance
Lightning Source LLC
LaVergne TN
LVHW091259150826
845673LV00006B/1468

* 9 7 9 8 9 9 6 3 8 1 9 0 6 *